Technophysiology, or How Technology Modifies the Self

Technophysiology, or How Technology Modifies the Self

By

Roberto Marchesini

Cambridge
Scholars
Publishing

Technophysiology, or How Technology Modifies the Self

By Roberto Marchesini

This book first published 2023

Cambridge Scholars Publishing

Lady Stephenson Library, Newcastle upon Tyne, NE6 2PA, UK

British Library Cataloguing in Publication Data
A catalogue record for this book is available from the British Library

ISBN (10): 1-5275-2882-0
ISBN (13): 978-1-5275-2882-6

TABLE OF CONTENTS

INTRODUCTION

The technological advancements of recent centuries demand a radical reevaluation of how we perceive *techne* in a philosophical framework. This requires moving beyond the humanistic model that views technology as a mere tool, enabling humans to enhance their actions, pursue their goals, be freed from their needs and weaknesses and exert control over the world. Such a conception of technical or technomediated action is built on assumptions that have proven untenable, both theoretically and in terms of practical application. Key pillars of the humanistic interpretation, such as the theory of human incompleteness, the ergonomic and disjointed nature of technological apparatuses, the compensatory, amplifying, and liberating aspects of technomediation, the autarchic understanding of the ideational and design process, and the assumption of complete human control over the functions and structure of machines, have all been discredited.

What emerges today is a new understanding of the organic. It is now viewed as a conjugable entity capable of functioning with external supports to generate performative outcomes, as seen in the theory of the extended mind. At the same time, it is also seen as a metamorphic entity capable of adapting and assuming new functions through the use of instruments that perform similar tasks and therefore replace the dedicated organ. The body's involvement with technology, from a physiological perspective, exceeds traditional conceptions. Hence the need to talk about *techno-physiology*. In an eco-ontological framework that builds upon the classical notion of interpersonality but extends it to a broader ecological conception, intentionality transcends the realm of consciousness and the human sphere; instead, it characterizes a hybrid ontological condition that emerges from our relationship with nature and technology. Indeed, technology itself transcends traditional boundaries, as it directly enters the body through deep interfaces: it has an organic constitution ("wet" or organic technology), possesses self-programming capabilities (deep learning), and assumes performative forms that are indistinguishable from those attributed to human beings.

We are witnessing profound shifts in the conception of *techne* that have implications for the ontological understanding of technomediation, the relationship between technology and the body with regards to induced physiological transformations, its social and individual impacts, future scenarios, and ethical considerations. Biotechnology has erased the distinction between invention and discovery, genetic algorithms have surpassed traditional notions of programming, biosynthetic grafts have challenged the concept of the body-technology relationship, and digital immersion not only goes beyond analogue reality but also risks disconnecting individuals from the real world. The direct connection of sensory systems with technological interfaces reshapes the principle of interaction between the subject and the world, transforming simulation into lived experience and ultimately shaping one's biography. However, this process does not lead to a loss of the corporeal dimension, as some might suggest. On the contrary, it results in hyper-somatisation, eliminating the need for interfaces and potentially rendering sensory transduction obsolete, ultimately questioning the very notion of individual.

The convergence of various technopoietic directions, including nanotechnology, biosynthesis, and neural networks, leads to a rapid acceleration of circumstances that surpass the capacity of traditional philosophies, rooted in humanism, to provide an adequate interpretation and predictive framework. There appears to be a significant gap, both in terms of timing and understanding, between what technology enables and the philosophical comprehension of the actual consequences and existential possibilities offered by these new devices. Embracing the concept of technophysiology means acknowledging the existence of an extended body that transcends the individual and is no longer solely under the latter's control. Technological progress is accelerating at a pace that is hard to fully grasp, but what is even more concerning, in my opinion, is that we still perceive it as separate from our bodies and external to our ontology. We continue to view technology as a tool, something to be wielded like a hammer, fully at our disposal, perhaps wearable like clothing, capable of serving our needs or competing with us, but still we persist in considering it external. This is a fundamental mistake.

Technophysiology instead urges us to adopt an integrative and non-dichotomous perspective, recognising that the niche in which an organism exists is an intrinsic part of its body rather than a separate habitat to be occupied. Just as a spider's web is an integral part of its phenotype, the constitution of our own physiology is inherently heteronomous, always relying on external mediums. Failing to recognize this interconnectedness is a significant mistake, one that today may prove more perilous than we currently realize. The cultural imagination still relies on rhetorical devices that hinder our ability to fully grasp the profound implications and transformative effects of technological innovations. Thcsc advancements, in fact, not only open up new existential dimensions but also create novel fields of research and application that were previously unimaginable. Today, the task of understanding technology is mostly carried out by science fiction, albeit inadequately. While this genre can offer some accurate predictions from a technological standpoint, it falls short in addressing the ontopoietic and existential dimensions of the matter. Indeed, science fiction remains a transreal projection that explores the human condition in a limited manner, unable to fully comprehend and depict the profound implications of technological advancements.

Allow me to elaborate. Science fiction, through its transreal narrative approach, explores how human beings would navigate and experience a future world shaped by technological advancements, geopolitical shifts, and the consequences of the ecological crisis. It projects contemporary individuals, with their unique perspectives and worldviews, into these altered environments. However, transrealism primarily serves as a tool for *our own* understanding and contemplation rather than a prefigurative representation. It allows us to confront potential transformations and imagine our responses to various challenges. Science fiction, particularly the works inspired by Philip K. Dick, has a psychotropic and psychoanalytic impact as it delves into the individual psyche, influenced by cultural and generational contexts, and places it within a transformed setting. Nevertheless, science fiction cannot replace philosophical inquiry as it cannot fully address the ontological and epistemological aspects related to change.

Technology has always brought about profound shifts in our ontology, altering the very essence of what it means to be human. Its impact extends beyond providing tools for better understanding, interpretation, and fulfilment of our needs as stable and universal beings. Transrealist science fiction, in this context, can only offer a moment of reflection on our existential condition rather than predict realistic scenarios. To comprehend the true implications of a radical physiological revolution, such as the one that is unfolding in our time, we must recognize that the emerging humans will have a different physiology and a fundamentally distinct mindset from our own. Consequently, their aspirations, goals, values, moral behaviour, dreams, fears, worldview, concept of time, memory, and thinking will all be transformed. This does not imply a complete departure from the human phylogenetic essence, of course, but rather a novel manifestation of our species-specific heritage. Therefore, embracing the concept of technophysiology does not negate the taxonomic specificity of the human being; instead, it is crucial for understanding the diverse possibilities that arise from our evolving nature.

The present essay aims to explore the distinct characteristics of the morpho-functional and ethological constitution of the human being in relation to technological integration. It emphasizes that the body possesses a certain virtuality or malleability but is not entirely fluid: somatic organisation continually adjusts and adapts to external supports, engaging in reciprocal processes of accommodation. With the advent of a new technomediated condition, performance standards and feedback mechanisms for various organs are altered. Through technological devices, the body develops new interfaces for perception and action in relation to external reality, leading to a transformation of sensory correction mechanisms, sensibility, and agency. This profound change in the organism occurs during the assimilation of the tool, affecting neurobiological connections, endocrine regulation, immune responses, sexuality, the musculoskeletal structure, organ functions, biorhythm, and cell regeneration. Hence, I argue that technology is always infiltrative, becoming embodied even when it initially appears to merely dress the body like a glove that protects it from contamination.

Today's technologies challenge us to move beyond our traditional understanding of devices and to embrace a new *philosophy of technology* that acknowledges the ontological transformation brought about by our relationship with these devices. To comprehend this transformation, it is essential to shed light on the somatic changes that occur when we incorporate new devices into our daily lives. This essay aims to examine the effects of technology on the body, focusing on its morphological configuration and the organisation of its functions. I use the term "technophysiology" to describe the profound impact that technology has on how we exist in the world – our perception of reality, the construction of our identity, our expectations, and the meaning we attribute to our lives. Being a body means living in coherence with the somatic dimension that defines us, but this dimension is not exempt from the modifications brought by technology.

CHAPTER 1

ON HUMAN NATURE

Premise

It is impossible to imagine technology outside of a close relationship with the body. Even tools that may seem distant from the bodily dimension, such as a traffic light or a hydroelectric turbine, ultimately require somatic transactions. They do so through the need for management techniques or attentional mechanisms, either directly or through modifications of the environment, by means of elective stimulation or encryption (Virilio, 1995). There is no doubt, therefore, that technologies consistently influence the biological matrix that constitutes us. *Techne* imposes a formative seminar on the body, which always has to plastically adapt to technology, no matter how much effort is put into finding ergonomic solutions. The body is also called upon to take full advantage of the performative dimensions inaugurated by each device in its specificity of use and in the feedback it produces. In short, the body is subjected to different types of pressure from technological equipment. Each tool, while being specifically addressed to certain parts of the somatic ontogeny, always acts systemically, like an ecological niche (Odling-Smee, 1988). Just as a cobweb is a spider's somatic measure, its extended phenotype, so the technosphere dictates the ordering coordinates of our cell populations.

The spread of personal computers, for example, in just forty years has profoundly changed people's lifestyles and, consequently, the cognitive functions used more or less frequently by the average human (Greenfield, 2015). As a result, people's posture, attentional system, arousal and stress mechanisms have changed as well. Thanks to the progressive miniaturisation and versatility of these media, computer technologies have quickly found their way into electronic games, mobile phones, and music players, becoming the gadget *par excellence*. Caught in such an embrace, the body has made way for the computer: as in a successful transplant, a graft has occurred that has altered the way individuals think as well as their social expectations. The increasingly immersive digital social interactions, so assiduous as to occupy the younger generations for more than eight hours a day, has had an unsettling effect on their psychological condition. The digital experience has become embodied in people's ways of being, as individuals have gradually realised that they are being watched on the Web and that they are leaving substantial traces of themselves in what were previously innocent forays into the inertial dimension of the analogue (Ferraris, 2020).

The immersion, interaction and reciprocation of the digital produces a very strong attraction as well as a psychotropic effect, so that people are in danger of sidelining direct relations with others, concrete multi-sensory contact with reality, the kinetic dimension of experience, and rootedness in nature. This undoubtedly produces a sentimental diseducation, so to speak, accentuating forms of egocentrism and narcissism (Lasch, 1979), as well as a loss of connection with the world, but at the same time it alters the developmental and organisational guidelines of the entire organism. Especially in the case of minors, who are going through their developmental years – when the use factor has a preponderant influence – the repercussions are considerable. Today we are already beginning to notice the – not always pleasant – consequences of the entrenchment and infiltration of these technologies. But no wonder! The body is a plastic entity that, though predisposed to a certain adaptive niche through phylogenetic selection, provides for *a-posteriori* variation, i.e. ontogenetic adaptation. This already occurs with the translation of genetic contents into the embryo-foetal phenotype: we know, for example, that certain substances ingested by the mother are able to influence the endocrine, immune and neurobiological structure of the foetus (Soto et al, 1991).

The technological apparatus acts as a niche which accommodates the body but at the same time imposes its morpho-poietic guidelines on it. As such, it should be considered in its influence on bodily development and overall organisation (West-Eberhard, 2003). This influence concerns *both* anatomical and functional aspects, i.e. the conformation that the body assumes, *and* the metaplastic and neoplastic predispositions, i.e., what the body can expect from the future with reference to the stressogenic, deficient, mutagenic and traumatic events that are expected to occur. Today, thanks to research in the Eco-Devo field, we know that a genome always implies polyphenies, i.e., phenotypic potentialities, a sort of starting virtuality from which the developmental environment extracts one of several possibilities (Gilbert and Epel, 2015). Therefore, one cannot ignore the weight that the living environment plays in the structuring of the body and, consequently, one cannot underestimate the technological pervasiveness in which the individual is immersed even before being born. This means that it is impossible to talk about technology without clarifying the nature of thc body and the way it behaves in interaction. The traditional ergonomic view does not grasp the mutual convergence of the two terms, because it has an essentialist, predetermined and static view of the body's morpho-functional architecture.

Somatic organisation always has to accommodate and tune in to the media it is exposed to, through reciprocal adjustment processes involving both the development of peculiar techniques of use and the advent of new developmental pressures on the organism (Cregan-Reid, 2018). It is an illusion to think that the body remains untouched by the encounter with technological devices, because it is characteristic of the phenomenology of life to rely on the world’s acceptance. The body is an entity that, while maintaining its basic structure, continually modifies itself and does so by following the pressures that come to it from the external milieu. It does not lack a form of its own, it does not have a fluid consistency such as to assume the shape of its vessel, but neither is it imprisoned within a static form. It is a *metastable* entity (Simondon 2020): i) it is endowed with a precise range of phenotypic variability, so that morphogenesis must still respect a certain perimeter; (ii) it has dynamic configurations, i.e., it is constantly changing, while retaining its basic identity; (iii) it is recognisable not because it possesses an essential quality (which never changes), but because it has state singularities that can be categorised by family. The body is a perspective entity and in this sense it is projected into the world.

So, when a technomediated condition is inaugurated, the existential perspective changes: what is altered is the image of the self in the world (Floridi, 2014), and thus not only the landscape of agency, understood as the set of possibilities available to the individual – both in the standards of performance and in terms of desirable goals – but also the sense of one’s own body in different situations. Through technological devices, the body builds new interfaces with external reality, so we can see a transformation of sensory correction mechanisms in the areas of both sensibility and agency. Each technological innovation, as it were, opens a new window onto the world and sometimes leads others to obsolescence or neglect, so we can speak of a true existential metamorphosis. On the other hand, not only does the perspective projection onto the world change: we could say that the landscape, the way of looking at it, but also the temporal frame, that is, the alternation of events and thus their configurations in the *continuum*, change as well. Digital technologies, for example, increase the speed of perception, so it is as if time contracted and some details of the landscape were lost, yet the space covered has increased (Rosa, 2010). This is the same experiential difference as the one that occurs if a given route is taken on foot, by bicycle or by car.

In the niche of the technosphere, the body is not simply housed; it does not rest in it, maintaining its purity and integrity as if enveloped by foetal membranes, but is infected by numerous viral agents that transform its internal metabolism (Leonhard, 2016). The body is infiltrated by technology, which penetrates it through a vast network of capillaries, reprogramming its internal organisation. Therefore, it is the entire organism that undergoes a profound change in the acquisition of a new technology. What occurs is a metamorphosis that involves the neurobiological wiring, the endocrine setting, the immune response, sexuality, the musculo-skeletal chassis, the main organ functions, as well as the biorhythm and cell turnover. This is why I say that technology is always infiltrative: it is embodied, even when at first glance it would appear to only "dress" the body, wrapping it like a glove, preserving it from contamination (Capucci, 1994).

1.1 The plastic and changing condition of the body

The body is always changing in relation to what surrounds it; this is why it is important to consider each interacting element, especially if ongoing, as an *ontogenetic factor*. There is no doubt that technology represents an important counterpart in people's lives, capable of characterising children's development matrix and the subsequent stages of refining and maintaining that morpho-functional configuration that can be defined in terms of individual singularity (Arthur, 2009). On the other hand, in order to understand the influence exerted by technology on somatic physiology, it is essential to shed light on the very concept of the body's plasticity. In fact, we might think that this plasticity arises from a lack of an inherent (proper) form or from incompleteness, whereby the technical or cultural element would merely compensate for the lack of equipment received from nature – like a vessel which, the emptier it is, the greater the amount of liquid it can contain. This explanatory model, which can be defined as the *scarcity principle*, has characterised the humanistic tradition since the classical heritage and still has its appeal, but it has proven to be totally inadequate to explain the body's adaptive capacities (Gehlen, 1988).

If we observe the structure of the various apparatuses that make up the human organism, we note that their capacity to adapt does not arise as a compensation for some deficiency, but rather as pruning of redundancies and in general as differential growth of the resources present. The neurobiological system, for example, presents itself at the beginning of experience with a *redundancy* of neuronal elements and synaptic connections that are then reduced through processes of apoptosis. This pruning is supported and directed by the different functional activations of the various *neuronal assemblies* of the connectome (Hebb, 1949): therefore, it is the relationship with the external environment that configures the physiological and, consequently, performative structure that the neurobiological network will assume. The subject's experience, especially in the developmental age, when the plasticity-virtuality of the system is at its peak, will define the psychological-behavioural identity of the individual. Moreover, the activation of a particular pathway in the network strengthens this connection, increasing the thickness of the myelin sheath, the capacity of presynaptic neuromodulators and the sensitivity of postsynaptic receptors.

The evolution of very different cognitive identities in humans is thus not attributable to incompleteness or deficiency, but rather to an exorbitant number of possibilities that allow for the structuring of a wiring singularity through i) the pruning of network redundancy and ii) the differential development of possible connections. If we consider other apparatuses, such as the immune system, we will see that the configuration it achieves, for example through vaccination, is realised through the differential replication of certain lymphocyte families that are rewarded, so to speak, by the neutralising efficacy of a certain antigen (Edelman, 1987). The muscular apparatus also has its redundancies that are set in an athletic profile through the law of use and disuse. In the case of the endocrine system we are faced with homeodynamic mechanisms of feedback interaction that manifest a plurality of possible equilibria. And in general, even the phenotypic translation mechanism extracts a configuration from the potentials present in the genome (polyphenism) through epigenetic silencing or emphasising mechanisms.

It is therefore necessary to move from an intuitive view of plasticity, based on the act of filling a gap, to a potentially counterintuitive model based on the *differential wiring of a complex system* (Zhok, 2011). Plasticity is, therefore, the ability to express one's potential by producing a peculiar configuration that reflects the boundary conditions without betraying its own intrinsic characteristics. Let us take an example: the foliage of an oak tree takes on a peculiar form based on the light conditions present, while remaining true to its species-specific traits. In the phenomenology of life, plasticity does not come from a lack of form but from the generative potential of form (Henry, 2008). We should therefore adopt a different model, which can be defined as the *redundancy principle*: the greater the complexity of a system, the more relevant its virtual condition, i.e. the range of expressive possibilities and configurations it can assume. The body is not adaptive because it is deficient, but because its redundancy of actualisation potential allows it to introject information from the outside world. Plasticity, therefore, does not derive from an absence of implicit contents, but from their complexity, such that it can reflect its own adaptive history through configurations in the making.

The organism necessarily adapts to the environment, and it does so not by producing a form from scratch, but rather by utilising the endowments received from phylogeny (inherencies), which in turn are the result of an initial form of adaptation, traceable to natural selection (Lorenz, 1973). In essence, the ontogenetic adaptation process, which produces our particular identity, takes place by virtue of a generative dialogue between innate characteristics and the environmental factors that the system will encounter in the course of its development. This is why it is essential to bring plasticity back to the surplus of intrinsic properties and the *singularity of outcome* to experiential occurrences. I speak of singularity because the actualisation process of a system endowed with high virtuality from the start inevitably presents a degree of unpredictability. This can be traced back to chaotic dynamics, as well as to the uniqueness of the configurative conditions. So, the identity of a body is not only unique but also unrepeatable: a replica would not be possible even if the system were subjected to the same adaptive sequences, as utopian as this claim may be.

I mentioned experiential occurrences because the flow of influences that alternate in the configurative pathway of a body has its own irreducible randomness, so that every developmental process is always a somewhat indeterministic event. Having said this, it is evident that the ontogenetic system, alongside an openness to the unpredictable flow of random events, also presents mechanisms to mitigate the variability of this randomness. We can speak of *strange attractors* (Lorenz, 1963) that, while not defining deterministic outcome trajectories, nonetheless establish field probability indices. In the mammalian world, for example, parental care not only serves to nurture and protect the cub, but imposes educational paths that assign a certain direction to its development, through modelling and the masterful intervention of reinforcement and inhibition. The adaptive niche of a species can also be considered an attractor of ontogenetic outcomes, because it lowers the variability of the environment and the randomness of occurrences, creating a specific growth milieu (Lewontin, 1983). In this sense, even the culture and technology of a certain social setting can be defined in terms of an ontogenetic niche and thus a strange attractor. To sum up: plasticity is given by the system's redundancy and its intrinsic virtuality, while singularity is played out between factors of random occurrence and the influence of strange attractors.

At this point, it is necessary to understand what kind of relationship is established between the inherent contents of a body, which establish the *gradient of virtuality* – the breadth and characteristics of the field of ontogenetic possibilities that it implies – and the external influences called upon to translate this virtuality into a form through *actualisation*. Today, it is clear that the contribution of both factors is indispensable in ontogenesis, but how should we understand their relationship? Once we have assumed the co-factorality of inherent content and external influences, it is essential to define the terms of this relationship. Once again, we must endeavour to abandon the easy suggestions of intuition, which would lead us to consider this convergence as a compensation whereby the external would make up for inherent shortcomings.

Indeed, if this relationship were complementary – with each factor filling the other's gaps – we would fall into the deficiency model, whereby the ontogenetic plasticity of the human being would be attributable to its incomplete nature. This is the explanation that has characterised the *humanist paradigm* (Pico della Mirandola, 1942), to such an extent that to question it is tantamount to ushering a crisis into humanism itself. To consider the human body as incomplete from a biological point of view is, as we shall see, to mistake the *consequences* of technomediation for the *cause* of the process of technological endowment. This is an explanatory inversion based on a logical fallacy, which is nevertheless consistent with an intuitive vision as well as useful for anthropocentric purposes of ontological differentiation of the human from the animal condition (Heidegger, 2001). The redundancy model departs from this in that it modifies the very conditions of the relationship: it does not posit an inverse proportionality between the two terms, but rather a *direct* one, whereby the relationship is dimensional. As a result, the plasticity gradient is directly proportional to the complexity of the body system.

If we move away from the Platonic metaphor of the empty vessel and rather think of a marble slab to be sculpted or a network to be configured, we realise that explaining plasticity through deficiency is nothing more than a gross miscalculation. Variety in configurability belongs to a complex system, not an insufficient one. Only a redundant, non-deficient structure can develop a high gradient of virtuality, with many possible variation paths at the beginning of the development process. In other words, only a marked richness of inherence is capable of presenting a substantial level of plasticity. At the same time, we must return to the concept of experience, because the configurability of the body is never its passive exposure to the external configurative factor. Redundancy produces a virtuality that on the one hand requires and enables the experiential moment, and on the other is responsible for triggering it: redundancy gives not only virtuality, but also vitality to the system (Bocchi and Ceruti, 2007). Redundancy, in fact, defines planes of complexity that must necessarily rely on external contributions to be realised, just as a ladder is needed to reach a greater height. That is, it creates heteronomy: a level of organisation that can only be maintained by virtue of a high negentropic and dissipative potential, both in terms of energy and in terms of information content.

Heteronomy is an implicit externalisation, a reliance on content from the world, a dependence on external supports to reach a higher level of organisational complexity (Marchesini, 2022). Heteronomy can only be realised through redundancy, the capacity to accommodate and introject the boundary conditions that the organism finds in the environment. Heteronomy is, therefore, an assumption: the wings of an albatross speak of thermals and air resistance, the shape of a dolphin captures the viscosity gradients of water, just as the gills of a fish expect a liquid world to be sifted through and the size of an insect testifies to the oxygen levels in the atmosphere. It is therefore impossible to understand the body through internal inspection alone, because every bodily structure always implies the world: in practice, it presupposes it. The body is like an incomprehensible text without notational references, and these are given by the co-factors present in the reality surrounding it (Scheler, 1970). Heteronomy implies experience, as an event that is not accidental but already predefined, since the somatic dimension is the emergent product of the world itself: the body is coherent with the world because it has been shaped by it.

Therefore, we have to view plasticity not only as a configurative possibility by means of external co-factors, but also as a *yearning for the world.* Plasticity is an implicit reference to presences that, in their chaotic-contingent fluctuation and conversely in their coherence and consubstantiality with the body, allow for individual singularity. Plasticity is thus also a way of looking at the world with interest, whereby the body is an exercise in extroversion, projection and hospitality. In this sense, it is not possible to speak of the body without referring to this openness to the historical dimension of occurrences that we call experience. In it, the body is realised in a proactive manner that is only apparently expositional, because it is precisely through inherencies that the body can *take a position* in the world (Deleuze, 1994). Experience is the capacity for listening, convergence and correlation, but it is equally the yearning for the world and the ability to create worlds. This dialogical reciprocity of experience makes the becoming of the body an unpredictable event because, just as in a dialogue, the outcome of the discourse is always open. We must therefore move away from all forms of determinism, both internal and external.

The body is an experiential entity, in that it is realised by overcoming itself: it accomplishes transcendence by projecting it into immanence (Tymieniecka, 2007). The plasticity it derives from redundancy makes configurability possible but at the same time exerts it through experiential yearning: the continuous hybridisation with external entities translates virtuality into innovative morphological singularities. In this sense, the body can be defined as a creative workshop. Experience is a dialogue with the world from which we emerge renewed and contaminated; however, this dialogue does not occur between two impermeable and extraneous entities. There are countless reasons for this, but I will only mention two. The first is that the inherent contents of the *body system* speak of the world and are consistent with it, as they have been shaped by the world itself: although *a-priori*, they are nonetheless *a-posteriori* compared to phylogenetic experience (Campbell, 1974). The second is that experience itself, from the earliest moments after conception, produces the assumption of external contents that become inherent to the subject, characterising its subsequent experience. Each interactive moment modifies the individual's internal contents, involving the introjection of many environmental elements, so that it is no longer a dialogue between innate and environment, but between the *past* – what has matured from the previous relationship with the world – and external factors. The environment, consequently, also changes as the interaction proceeds.

As mentioned above, plasticity lies in the willingness to let oneself be "translated" by the external world, by the intrinsic virtuality that is capable of mirroring ontogenetic boundary conditions: through the mechanisms of pruning and evolutionary differential, the body is able to introject information (Changeux, 1986). Plasticity is also given by the propensity to do so, in what I have called *experiential yearning*. The expressive process is not a passive event, where the world is left to select the variants acting on a body that is simply exposed to external influences. Mirroring is a phenomenon of absolute protagonism, where the body plays an active role. Experience is realised not only for the trivial reason that there is a world full of evocations and activities, but because the body is interested in them. The process of morpho-poietic singularity is therefore based on a yearning for the world that creates involvement – that is, openness and interest in the outside world – and that is also of central importance in the way we approach technology.

Somatic plasticity generates experiential vitality and, from this, the evolutionary creativity emphasised by Henri Bergson. According to the latter, the body is always the emergence of a singular perspective on the world (Bergson, 1907). This implies an interactivity that underlies the phenomenology of life and that cannot be assimilated to other virtual but inert structures, such as the previous example of the marble slab compared to the statue. Life is expansion, infection, colonisation, replication, active pursuit of possibilities, hunger for energy, occupation of space, will to power and power of the will. Plasticity means taking an interest in the world as bearers of inherent drives, whereby feeling and desiring are motivating foundations. Experience is a probing of the possibilities of reality as holders of an intrinsic motivation, which drives the body to interaction through imaginations, hypotheses, transfigurations, epiphanies, and questions. We can say, then, that the body is plastic because it is involved, because it is not disinterested, because it is involved by the world and is therefore sensitive to calls, solicitations, challenges and elicitations. The proactivity of the body is not, in essence, a self-referential motion, but rather openness and self-giving (Marion and Tardivel, 2018).

The body can be envisioned as an exclusively dissipative entity, absorbing energy from its surroundings, but this interpretation, though true in some ways, is incomplete in others. The body, with its plasticity given by its virtuality and its involveability, is a creative entity that constantly produces singularities through processes of dialogic convergence (Haraway, 2016). On the other hand, if experience is not passive exposure but a dialogue, akin to asking questions about the world and formulating epistemic innovations through hypothetical-deductive paths, it is clearly not possible to do so without presuppositions. There would be no plasticity without an *a-priori* platform: thus we should not think of inherence as obstructive to the process of ontogenetic singularity, but as a constructive factor in it. Its role is to catalyse and trigger experience, and likewise to provide it with those basic endowments that enable its implementation. If we think of the body as an incomplete entity that seeks a sort of crutch in the world to find an adaptive centre of gravity, we fail to understand the primary meaning of experience which, conversely, is based on conjugation (Piaget, 1970). We must therefore pre-emptively ask ourselves what *copulative factors* act as a magnet for the world, creating a gravitation that is never thrownness.

We recognise the plasticity of the body by observing its expressive path; that is, we evaluate it on the basis of its capacity to assume multiple configurations, to reflect boundary conditions and to produce singularities. If virtuality were not functional to these outcomes, it could be dismissed as a simple amorphous, not really plastic condition. On the contrary, however, virtuality is plastic because it associates availability with experiential involvement: it proactively gives itself to the relationship with the world, as if intrinsically attracted to it (Patocka, 2016). We must therefore start precisely from this relational fascination that characterises the phenomenology of life: the implicit reference to an attribution of meaning. A child's gaze is full of astonishment and yearning towards the world, his smile in front of his mother's face is evidence of the joy of finding himself again; it is certainly not a confused bewilderment at the strangeness of his surroundings (Rosmini, 2009). The prevalence of a rationalist conception of somatic experience causes a neglect of affective factors that, on the contrary, are a prerequisite for interest in external reality. Technology also produces its ontogenetic influences by acting on the human emotional and motivational factors that create involvement in the immersive experience. Plasticity is commensurate with the gradient of openness, we might say wonder, amazement, exploratory pleasure and desire for self-assertion – all factors that accentuate immersion by acting on interest and involvement, and thus on the depth of experience.

Here, too, we can see how the influence of the external milieu cannot be attributed to a generic vacuity of the body, but rather to its copulative nature, which we can summarise in the body's readiness to be involved in experiential possibilities (Lichtenberg, 1989). The technological factor exerts an appeal that cannot be trivialised as a mere performative need, because it solicits the subject's intrinsic emotional and motivational systems: in other words, technology can speak directly to people's hearts and guts, even before their reason. Ontogenetic plasticity involves experience, a process of integrating environmental elements that translates the initial virtual condition into a specific form. But experience is not a simple exposure to the world: it is an immersive process that requires involvement on the part of the subject.

This involvement that makes the world attractive to the subject's eyes is given by affectivity, in its emotional and motivational characters. Our sentimental and desiring nature plays a fundamental role in our technological experience, as well as in all those processes that reveal plasticity *a posteriori* (Pulcini, 2012). We must not forget that the body is affectively copulative because it evolved through yearning: (i) it has specific metabolic needs, such as hunger and thirst; (ii) it has a certain hedonic predisposition, marked by an exteroceptive aesthetics and proprioceptive pleasure; (iii) it assigns a dispositional, rather than rational, value to events. This makes the subject interested in the world, not detached from it, and thus inclined to openness: it tends to welcome what is outside and to project itself into the world.

I have defined experience as a dialogue with the world, and this aspect recalls the inadequacy of the expositional conception and the deficiency principle, because a dialogue always refers back to the input of the interlocution. If the ontogenetic result is full-bodied in its vastness and variety of outcomes, as we can see in the human being, it is precisely thanks to the richness of the resources that precede experience itself and make it possible to formulate questions or hypotheses, giving rise to a singular perspective (Dehaene, 2020). It has to be said, however, that by experience I do not only mean the cognitive-behavioural aspect of interaction and explicit learning: muscles, glands, bones, and intestines also experience reality. Therefore, by inherent content I mean more than the classical transcendental *a-prioris*. For example: i) the sensory organs, as media, are contents to all intents and purposes; (ii) time frames construct a precise image of continuity in the configuration of experience; (iii) the enteric structure and intestinal symbiosis establish orientation contents; (iv) endocrine and immune systems define a biorhythm for the entire adaptive metabolism of the apparatuses.

The body is not closed in on itself, it is not autarchic and impermeable in its essential, monadic fixity, but has evolved by implicitly taking on the ordering action brought about by external entities. Heteronomy is, therefore, an intrinsic characteristic of the body that presupposes a contribution beyond its own individual presence. I said that an organism could not maintain its internal organisation without energy and information from the external environment: in this sense we can say that it is an open system (Prigogine and Stengers, 1979). At the same time, the body could not subsist without a hereditary content, that is, without the contribution of a package of information that comes to it from phylogeny. For this reason, we can speak of a *dissipative structure of experiences* that preceded it. In other words, the body is a diachronic entity: without the hereditary legacy, a living being could not open itself up to experience, without a niche it could not achieve adaptation, without an ecological-relational location it could not position itself in the world (Odum and Barrett, 2005). Plasticity, therefore, is based on heteronomy, i.e. on the presence of a somatic dimension whose realisation already implicitly envisages the contribution of the world. It is plastic insofar as it is not based on autonomy, autopoiesis and self-sufficiency. The presupposition of the world allows the body to reach an organisational level that would otherwise be impossible, so plasticity is the very ground of the *phenomenology of life*.

Plasticity is openness but not passivity, and this awareness also produces an explanatory reversal of no small importance in understanding technophysiology, which we would be wrong to view as a superstructure or as something that dominates the body. If it is incorrect to read technology with the classical focal point of the passive tool in the hands of the human being, it is equally misleading to imagine a body ruled by technology. Experience is a yearning for the world that involves the body and expresses it through the environmental matrix, whether it is the result of technology or not. Without involvement there is no experience and no expression: plasticity, therefore, is not fluidity or an amorphous condition, but the creative manifestation of inherencies in their morphogenetic capacity. It is therefore necessary to presuppose implicit contents that not only allow for plasticity, but also implement it. Without affective components, for example, there would be no interest in the world and, more generally, there would be no involvement that transforms experience into a moment of immersion (LeDoux, 1996). Likewise, without *a-prioris* there could be no experience, because we would have no questions to ask and no hypotheses to formulate. The plasticity of the body does not lie in the absence of inherencies or intrinsic endowments, but in the experiential vocation – that is, in the heteronomy of the system itself.

1.2 The somatic dimension of technological experience

Looking at the relationship between body and technology is a necessary step in understanding the technical dimension of human experience and how the somatic dimension as a whole is strongly influenced by the technosphere. How is it that our species has developed such an articulated and profound technical dimension that has no equal in nature? This technological propension substantiates the Promethean definition that characterised classicism. As mentioned, the answer given by the humanistic tradition is based on the deficiency principle, assuming that cultural intervention in its most generic sense – and therefore including the technical sphere – should be understood as a compensatory factor (Herder, 1869). Instead, I will try to set out a different line of reasoning, no longer based on the supposed incompleteness of human nature – that is, of some original deficit to be compensated for – but by investigating the characteristics of our species' phylogenetic heritage that have enabled and facilitated the emergence of such a robust, articulate and multifaceted cultural dimension in human history. I will therefore not attribute the technical emphasis of the human being to a lack to be made up for through the use of tools, understood as an adaptive crutch, but to propensities and endowments that have evolved characterising the hominid lineage (Leakey, 1994).

We know that mammals also have cultural traditions and that primates, in particular, use a plurality of tools. Chimpanzees, for example, use sticks as weapons against potential enemies, fashion longer sticks to catch termites, use stones to open nuts, and use medicinal plants to cure diseases (Goodall, 1990). Lithic art was already present in the Australopithecines and developed progressively with the advent of the genus *Homo*, over the course of 2.5 million years, until it reached extremely complex forms in the Aurignacian period, some 40,000 years ago. Therefore, we must assume an already established tendency in the *Hominidae* family, which has become specialised in the genus *Homo* through processes of adaptive selection. This means that, contrary to the customary interpretation of human nature and, consequently, of the technical tendency, this propensity must be sought precisely in *coherence* with the phylogenetic heritage of our species. Obviously, when I speak of propensity, I am referring to certain qualities that have fostered technical expression and others that are the very outcome of technological behaviour, in a process that can be described as recursive.

The technical dimension can be represented using the theory of niche construction, which involves: i) on the one hand, a species already predisposed to modifying its environment through particular operations – think, for example, of the beavers' dam building; ii) on the other, selective pressures directed or modulated by the niche itself, defining a particular phylogenetic trajectory and increasingly tying the species to its niche. In other words, we can identify in the genus *Homo* a technopoietic emergence of a character already present in primates in general and in the *Hominidae* in particular. This strengthening may have been stimulated by environmental adaptive factors, but also by changes in population. The increased use of tools would then have contributed to modifying selective pressures or Evo-Devo effects on hominids with recursive repercussions acting as a driving force for further technopoietic developments. It is therefore necessary to see human nature not as something incomplete that requires tools to compensate for its maladaptation, but as full of productive, dispositional and creative traits underlying its technopoietic tendencies.

The expressive plasticity attributable, for example, to neuronal redundancy may be a good starting point for understanding both the gradient of niche predisposition in the technical sense and the recursive effects of technomediation. There is no doubt that human beings have a pronounced tendency to modify their environment, most likely because their phylogenetic cradle presents strong state fluctuations and habitat characteristics of considerable problematicity (Morris, 1967). Here, too, plasticity proves to be a source of adaptability and likewise of acute experiential singularity: it is a still recognisable quality that causes human beings to assume identity profiles reflecting their circumstances of growth. In other words, the influence of technology fits into a natural human condition, a trait already predisposed to variation by external factors. The biological characteristics of our species – involving neonatal immaturity, a strong motivational drive, a highly developed neocortex, and the social dimension – are an inescapable prerequisite when it comes to understanding the technical predisposition and the ontogenetic influence exerted by technologies. Contrary to the claims of the humanistic tradition, there is a close and directly proportional relationship between bodily plasticity and phylogenetic heritage.

So, human nature is by no means deficient. It is precisely by virtue of its redundancy that it achieves its degree of virtuality, which gives it a high level of openness to circumstances and to the attainment of singularity. This is how we must view technology, rather than thinking of it as a crutch that compensates for supposed biological inadequacies. The technological dimension must be analysed in terms of an ecological niche: built on the basis of latent propensities in the phylogenetic lineage, in turn, it acts as a flywheel for stimulating such tendencies. As we shall see, the relationship with technology produces considerable influences on ontogenesis through various avenues, the most important of which is the differential exercise of organs and apparatuses. On the other hand, we would be equally mistaken if we considered the technosphere to be a kind of foetal membrane that shields us from external contamination (Sloterdijk, 2011). In reality, technology does not disconnect us from the world but intervenes by modulating its relationship with it, and in this sense we speak of technomediation. In some areas, there is a greater dependence on external contributions because the system assumes a more dissipative configuration and a higher level of integration.

We are wrong, therefore, whenever we depict the technological dimension as a containing envelope, a surrogate for a uterus, or as a sphere that preserves, protects and decontaminates us. Conversely, technomediation is comparable to an anastomosis that metaphorically expands our circulatory system beyond our body, placing it in even more direct contact with external reality and, above all, making it more in need of energy and information. It is therefore indispensable to talk about *technophysiology*, today more than ever. In fact, we are still caught up in an explanatory framework that considers technology as a garment that, while characterising us, does not alter the presumed purity and autonomy of the human being. On the contrary, in technological hybridisation, we do not divest ourselves of our human condition. We do not become inhuman: we realise dispositions that are inherent in our nature (Eibl-Eibesfeldt, 1989). When we imagine the technosphere as an exoskeleton that envelops us, standing alongside our biological condition, we make a big mistake that may even be the harbinger of great sorrow. In reality, technology dissolves any barrier between us and the world; if anything, it operates a somatic boundary reduction, making the body more permeable and unstable. Every technology bursts the body open, scattering it to the four winds, precisely because it reinforces physiological openings.

A technology can shield an organ from the strain of performing a certain function, while simultaneously subjecting it to a different performative pressure. Moreover, this process does not increase the autonomy of the body with respect to the outside world, but rather reinforces the dependency gradient. In fact, the more a system increases the devices needed to achieve a certain result, the more it requires negentropic mechanisms (Schrödinger, 1940). Our dependence on the world is not to be assessed in terms of mere contact, but in terms of the number of input levels it produces: a greater complexity of the niche inevitably results in increased energy demand. Moreover, a device is able to raise the performance bar and sometimes make otherwise unattainable goals possible. This is not to say that it absolves us from the contract with the outside world or that it diminishes the effort required, for the simple fact that raising the quality of the result pushes the body ever further to the limits of its powers. This is a fact for all to see: technology makes the body *more* needy and *more* projected towards high-profile goals, which necessarily require *more* external inputs (Gawdat, 2021).

The plasticity of the body and the readiness of human nature to open itself up to external contamination produce the metamorphosis, especially in physiological dynamics, that we derive from the different variations of cultural dimensioning, of which *techne* is an integral part. A sort of ontogenetic dance is thus created between body and technology that penetrates deeply into the metabolic meshwork and reshapes the somatic dimension of the human being. Today, we are witnessing a heated debate on the interactions between the body and technology, which can be traced back to the development of certain practices – I am referring in particular to the digital world and biosynthesis – that have literally revolutionised our perception of the human being's operational potential (Bostrom, 2014). On the one hand, these advances set off a fairground of futuristic fantasies; on the other, they raise perplexities and fears, which can be summed up in the question of whether everything that is technically possible is also licit or desirable. By a strange convergence of factors, as if in a space-time wormhole, the end of the 20th century brought to the fore the reflections of 15th century humanism, questioning the modern age.

This calls for a new philosophy of technology and, at the same time, a new anthropology, so as to break away from the traditional positions that have characterised humanism over the last five centuries. The challenge is not to give rise to new forms of anti-humanism – a sterile deconstruction that does not offer interpretative and propositional keys for the future – but to trigger a new paradigmatic revolution capable of offering a better picture of the human condition immersed in an increasingly complex and articulated technosphere (Benasayag, 2015). The somatic dimension is no longer interpreted as adherence to a stable form, but rather as a *generative principle of possible forms*. This, although confined within a range, presents a gradient of ontopoietic freedom that requires a much more articulated reflection than the Vitruvian canon. We can see this shift as the advent of a new awareness of the necessarily hybrid condition of human ontology, taking leave of the essentialist, autarchic and emanative approach so well represented by Leonardo's image. Becoming aware of the ontogenetic mutations occurring in the increasingly frenetic succession of new technologies requires a critical vision capable of avoiding both easy techno-enthusiasm and neo-Luddite pessimism.

At the centre of my discourse, therefore, is the somatic dimension of technical experience and, in particular, of the peculiar form called technomediation. From a merely external and instrumental conception of technological devices, as auxiliary entities called upon to passively serve the dictates of the human being, we must move on to a hybrid vision capable of profoundly transforming the somatic dimension of the human being in all its expressions. Instead of accepting the compensatory reading, which claims that *techne* acts as an external – therefore non-contaminating – crutch, I will question the somatic dimension in order to understand what biological factors have allowed our species to excel in the technical domain. If in the supplementary conception we turn to alleged somatic deficiencies to explain the exemption and compensation offered by technology, in this book I will attempt to investigate the phylogenetic specialisations that have enabled the technological emphasisation of the human niche. Our bodies have a certain plasticity, but it is equally clear that our cognitive endowments, our practognostic and eye-hand coordination skills, our social and collaborative propensity, and our abstract and representational intelligence have made the technical-cultural trajectory of humanity possible (Morin, 1974). It was not deficiencies that drove us to the use of tools, but rather a strongly copulative and creative psychic dimension, an extraordinary memory, and a body that is ductile in its kinesthetic abilities and manipulation of objects.

The effects of technology on the grammar of the body also require us to reflect on how technology manifests itself to the subject's experience by redefining the individual's life (Carr, 2010). The close interaction between body and technology does not only concern the performative aspect, but also the many facets of a person's life – things like social participation, wellbeing expectation, physical condition, or the expression of propensities. Therefore, we need to be aware of the strong influence of technology on the individual's existential dimensions (Vaidhyanathan, 2018). In order to understand the functional remodelling effect, however, it is necessary to overcome a number of misconceptions about the body and technology that still hinder the development of a truly interactive interpretation of the two terms. Accustomed to considering devices as tools to be used, in a unidirectional logic – as if we were immune to any induced consequences – we do not realise that every innovation produces a true somatic revolution in the body (McLuhan, 1964). The thesis I will illustrate in this essay is that each device should be considered as a body-building machine, capable of modelling the structural and functional organisation of the body.

This comparison may seem paradoxical and should certainly be taken as a simplification, aimed at illustrating the general mechanisms of *technophysiology*. The muscular conformation and the influences that exercise determines in its structure are obvious: a cyclist strengthens certain parts of the body, while a swimmer and a weightlifter develop others. Conversely, those who lead a sedentary life run the risk of muscle tone loss and, in cases of total inactivity, serious forms of atrophy. Even the body's posture, if repeated over time (e.g. someone who spends many hours on a mobile phone or computer) causes muscular tension that acts on the joint systems, intervertebral discs and cartilaginous portions (Cregan-Reid, 2018). The exercise of specific parts of the musculoskeletal system determines a growth differential and is easily observable in outer appearance. Other somatic dimensions are not as obvious – I am thinking, for example, of the neurobiological wiring or the endocrine setting – however, the differential development mechanism caused by exercise is the same, so that our brain and hormonal metabolism also receive growth or decrease impulses depending on how much the different apparatuses are stimulated (Siegel, 1999).

The somatic dimension is a plastic entity, a kind of ecosystem between competing and cooperating populations. The body is affected by a large number of constituent factors and systemic dynamics, achieving a condition of dynamic equilibrium between the forces in play. In this stage of climax, it is difficult to establish the boundary between internal and external, between stability and becoming, between intrinsic and extrinsic (West-Eberhard, 2003). The architectural and functional conformation is sensitive to environmental influences, especially during the developmental years: what happens is a genuine introjection of otherness. On the other hand, at every single moment there is an ongoing remodelling, however small, involving tissues and system assemblies. The external element may disarticulate or conjugate cellular groups, forming new functional convergences that reflect the contextual conditions (Waddington, 1959). Technology is thus never external, but always embodied, modifying the texture of the organism. This is nothing new in biology, where an entity is never viewed as an essence but always as a consortium. Suffice it to say that 90% of the cells that make up our organism do not have a human genome: they are symbionts that do not merely inhabit the body, but constitute it and define its functions (Margulis, 1991).

Plasticity makes the somatic structure susceptible to external influences. Technology is no exception, and it should come as no surprise if our physiology adapts to it, matching the various devices it engages with daily. In some ways, these devices have the same effect as nutrition, which, with its excesses and deficiencies, alters metabolism. The body needs nourishment and cannot maintain its thermodynamic complexity without receiving energy from the world: the external contribution is, therefore, the foundation of metabolism itself, the consequence of the co-evolutionary pact established in phylogeny. We are relational entities that nourish not only our energetic appetite, but our very ontology, by leaning on the world and introjecting its imprints (Vygotskij, 1962). Technology is a form of nourishment, something that enters into the meshes of our bodies and transforms them: sometimes so subtly that it is not easy to be aware of it, at other times blatantly, with a true metamorphosis affecting both morphology and behaviour. The organism, in fact, is an adaptive entity that presents a condition of phenotypic virtuality and, consequently, can give rise to different anatomical and physiological morphotypes (Mayr, 1963).

This plasticity or phenotypic virtuality must, however, be contextualised, assuming that having the capacity to adapt to external conditions does not mean lacking content or being passively exposed to the world. Even over different ages, the body maintains a degree of continuity, which is recognisable if one follows the evolution of a physiognomy. While it is true that identity is the permanence over time of a succession of events inscribed in the flesh, of somatised encounters and metabolised otherness, it is also undeniable that these acquisitions are positioned within a predefined framework. Stating that external factors influence the development of the body does not mean denying the existence of a species-specific canon, characterised by basic architectural lines (Carlson, 2013). The matrix of the organism has its own very precise structure, albeit plastic and influenced by external factors, and it is thanks to this constitution that the body is able to introject environmental influences, inscribing them within its identity (Abouheif et al., 2014). The somatic dimension is, therefore, the site of reception, a substratum endowed with its own conformation, capable of incorporating the incessant becoming of influences bound together by the very identity of the individual (Bateson and Gluckman, 2011).

The body is therefore diachronic and inclusive. Its plasticity is neither total liquidity (a condition that would lead it to always take the shape of its container), nor a lack of its own organisational matrix, which would condemn it to the determinism of the occasional and the contingent (Tinbergen, 1965). Rather, its plasticity grants the body a certain conformative freedom to move within a virtual space, becoming a malleable substratum, subject to the influence of the chaotic flow of possibilities that is readily introjected into its constitution. The body is thus a relational entity that faces the dialectic of encounter through the synthesis of somatisation. For this reason, concomitant factors have a considerable impact in shaping the body: we are also the otherness with which we engage (Levinas, 2000). It is not a matter of counterparts or space: a niche is the expression of an inclusive principle, where the becoming of the body is always the outcome of a dialogue. Any modification of this principle inevitably has repercussions on ontogenesis, because it involves the very course of the body-river, both in its configuration and in its evolvability. The capacity to introject organisational information is, in the final analysis, a quality of the body, which necessarily demands interpretation but already gives us the measure of incorporation.

It is clear, then, that reasoning about the somatic dimension of technical experience is a prerequisite for any reflection on technology and, more generally, on the kind of influence that cultural institutions exert on human nature (Remotti, 1996). How are we to interpret this dialectical propensity of the body and its capacity to follow relational trajectories? When we talk about technology, I always have the impression that we assume that it is an extension, as if the body operatively extended itself into the world while keeping the hand distinct from the tool. But the body's ecosystem is not clear-cut, so it is no use to claim to be able to extract its supposed essence in purity (the "true Self"), considering that even the hand risks turning into a tool by cognitive absolutisation (Goldberg, 2016). Continuing along these lines leads to an infinite regress that eventually generates the illusion of the information string, so it is necessary to look at the terms of identity co-factorality differently (Fredkin, 1990). The technosphere is certainly an essential factor in the human niche, an assumed dimension of the matrix, a grammar of the somatic narrative and the expression of the flesh. Here, the matter becomes more complex compared to the usual analytic way of considering technology: each invention does not merely add a new device to the technological aggregate, but modifies the niche in a systematic way (Noble, 1997).

Therefore, far from being a mere tool to be used by humanity, remaining outside the body and therefore disposable once its contingent necessity is over, the technological device changes the somatic dimension of experience. What we still refer to as a tool is actually much more than that, because it acts on the body's *development mechanisms* by redefining the weights and internal organisation of the system as a whole (Gould, 1977). We know that the body's plasticity allows for very different existential dimensions, which is why we cannot speak of technology as an external object that can be put away once we have finished using it. Even more so, we cannot think that technology merely enhances or extends the performance of the human being. In other words, we must not restrict its effects to performativity alone, because its effects on the body, while being much more relevant in the organs most involved in its use, always have a systemic value. All technology transforms the whole system of the organism, changing the experiential character of the body in its totality. Finally, it is necessary to remember that the modification of the developmental condition of a biological population also exerts effects on the phylogenetic trajectory, as shown by Evolutionary Developmental Biology (Raff e Kaufman, 1983). So, we should ask which characters in human nature have favoured technical experience and how the latter has shaped the somatic dimension of hominids over the last two million years.

1.3 Incompleteness or specialisation?

In order to understand how technology is capable of transforming the morpho-physiological matrix of our organism, it is essential to review certain descriptive and explanatory principles used by humanism to define human nature. The hypothesis of man's incompleteness and lack of biological specialisation, which characterises philosophical anthropology, gives technomediation an *exonerative mandate*, i.e. one of regulation, channelling, compensation and empowerment (Gehlen, 1988). Technology, in this approach, becomes a sort of crutch that supports/contains the somatic energy, surrogates its functions without, however, affecting the flesh. The body, impervious to external contamination, is enveloped by technology, which preserves its purity and its fidelity to its essence. Viewing human nature as a larval condition, undefined, immature, and lacking adaptive specialisations, i.e. incomplete and insufficient, becomes the axiom on which to build an anthropology of ontological distance from animality and of constitutive self-determination (Scheler, 1970). These two principles are the foundational prerequisites for the autarchic and autopoietic conception of the trajectory of humanity as well as for the hold of humanist assumptions.

As in all disguised tautologies, here too the system stands by virtue of explanatory epicycles (Kuhn, 1957). It is evident, at this point, that the way we interpret the somatic dimension – and human nature in general – radically influences the entire conception of the *technical dimension*. As long as we look at the biological condition of the human through the distorting lens of incompleteness and lack of adaptive specialisation – the interpretative key inaugurated by Pico della Mirandola – we can hardly go beyond the suppletive, compensatory and exonerating principle of the cultural apparatus. In humanist philosophy as a whole, the reduction of the biological factor becomes instrumental, I would even say necessary, to construct a human image fully emancipated from *bios* or at least on the way to becoming detached from the natural realm (Manetti, 1975). This human being sees all other species as a shore from which to distance itself, and the telluric as gravity – something to be abandoned in a rush towards the heavens, an analogy of the empyrean. *Techne*, but also every other cultural apparatus, thus becomes the armour forged by Prometheanism: it is what enables the ascent towards the angelic dimension. In this perspective, nature's bearing on the human being is reduced to define a boundary line useful for the realisation of *dignitas*, i.e. an ontopoietic route (Garin, 2008).

The humanist approach reads technology as a garment, an external element that rests on the body, compensating for its presumed deficiencies without, however, changing its predicates – on the contrary, preserving them in their purity. It is important to emphasise this last aspect: by viewing technology as a suit of armour to be worn, one not only avoids connecting the body to its function, maintaining its totipotence, but also separates corporeality from the impact that performance would entail. Technology would thus perform the same function as a glove, protecting the hand and ensuring non-contamination (Gehlen, 1988). This reading has undoubtedly long characterised the Western tradition, because it is consistent with a certain vision of the human condition, functional to sustain the very spirit of the modern age. However, since the 1990s, this dualist and disjunctive approach has been called into question, especially following the computer and the biosynthetic revolution. This was a veritable paradigm shift, inaugurated when the new technologies showed that the pretence of maintaining a difference of domain between the biological and the technological – between invention and discovery, between natural and artificial, between carbon and silicon – is no longer valid. Nevertheless, we still insist on an ergonomic interpretation of the technological device, in spite of the fact that technological grafts are increasingly complex, intelligent and of ordinary use (Langton, 1990).

On the other hand, the humanist focal point, which places the entire technical dimension outside of nature and makes it a compensatory support, can only bring every technological innovation, however infiltrative it may be, back to the same framework of explanation, on pain of undermining the entire paradigmatic structure (Plessner, 2019). In other words, a different culture of technology cannot evolve as long as we preserve the anthropological matrix which defines the human being as a biologically incomplete entity, one that turns to the *techne* to compensate for the lack of a precise adaptive rank. Ultimately, it is the very philosophical assumptions of humanism that anchor technical doing to a mere *suppletive principle*, hindering new lines of interpretation. This frame of thought does not take into account the hybridisation and metamorphosis of the somatic dimension of experience: it cannot do so because it considers the body the sacred temple of the soul and technology an amorphous entity external to the existential condition (Kelly, 2011). My impression, therefore, is that a new philosophy of technology is struggling to develop as a result of paradigmatic inconsistencies: as long as the biological incompleteness of the human being is presupposed, in a fully *a priori* sense, it will be up to technology to fill the gaps, to exempt us from tasks, to enhance and extend our performance.

One cannot speak of *techne* without referring to the *natural-cultural* condition of our species, i.e. without an anthropological theory (Haraway, 2016). In the humanist system, it is assumed that the human being is an uncontaminable entity that differs from all the other beings that make up the biosphere: while the latter are excellently equipped with biological means of adaptation – and precisely for this reason are circumscribed within a precise natural rank – the human being is seen as lacking biological endowments, incomplete from a natural point of view and therefore in need of a technical-cultural crutch. This condition, moreover, would make humankind alone free and self-determining with respect to its destiny (Scheler, 2009). It is evident that these assumptions place the *somato-technical* relationship in a compensatory and, at most, enhancing logic, light years away from taking into consideration any technophysiological repercussions. Reasoning about human nature is therefore indispensable for any reflection on technology: as long as we think of the human being as an entity that turns to technology to make up for biological deficiencies, we will never be able to understand how the body and technology converge in bringing out a precise somatic dimension of existence (Wolfram, 2002).

Questioning the deficiency principle as the cause of technical action leads to a different anthropology that profoundly revolutionises the body-*techne* relationship. If we cling to a compensatory reading, we are led to believe that technology develops by treating the gaps in the body as a filling mould, maintaining a mere juxtaposition to the somatic dimension. This means that it is the body that informs the predicates because, even in its deficiency, it nonetheless precedes them, so *techne* merely makes something possible that was already presupposed or extends something that was already there as a potential. On the contrary, we should consider technopoiesis as a creative and not a compensatory act, whose result is always emergent, i.e. never presupposed (Noble, 1997). To be even more precise, we must recognise that the consequence of this paradigm shift is even more radical, because it unhinges the traditional coordinates of ontology, assigning the predicates manifested by the entity not to inherent qualities, but to relational emergences. Overcoming the theory of incompleteness means inaugurating a profound ontological transformation, capable of revising the philosophical assumptions of technology in its ideational, existential and perspectival aspects (Chaitin, 2010).

But to address this line of reasoning, it is necessary to show in detail the distortions caused by the aetiology of deficiency. There are a few points that show the full explanatory precariousness of the compensatory theory. Let us examine some of them: (i) it is not possible to correlate a specific biological deficiency to every technological innovation; (ii) if deficiency were the ideational engine of technology we should hypothesise a trend of progressive decrease of technological innovation, whereas the exact opposite is true; (iii) if the process were compensatory or potentiative we would find a human being that is more and more stable, yet this is not the case; on the contrary, we note an increase in non-equilibrium; (iv) if technology were limited to operating a functional compensation there would be no deviations in the human condition and its teleological projection.

It is evident, then, that in order to speak of technophysiology, it is essential to make an accurate and objective examination of *human nature*, trying to understand the kind of relationship that is established between the body and technology. We can say, therefore, that it is indispensable to carry out an accurate analysis of the biological condition, in a way that is as objective as possible, i.e. free of prejudice or the need to prove a hypothesis at all costs. Obviously, it is neither the intent nor within the scope of this essay to address this issue in a comprehensive manner. I simply wish to show the obvious inconsistencies of the hypothesis of the incompleteness and lack of specialisation of *Homo sapiens* (Bolk, 1926). Our species is the result of a phylogeny that has subjected different human populations to selective pressures which have shaped their morpho-functional conformation. Among other things, primates are a mammalian order that has had a very specific adaptation, as evidenced by: (i) the five-toed structure on each paw, with an opposable thumb and short nails that facilitate a firm grip on branches and a prehensile ability; (ii) a vision with great colour discrimination and binocular vision, with the eyes facing forward and allowing a good depth of field as well as three-dimensionality; (iii) a vast food menu, given by a varied dentition, a wealth of taste receptors and a very versatile gastro-enteric structure.

If we then consider the morphological characteristics of the human being and compare them with those of other anthropomorphs, we immediately realise the high degree of specialisation achieved by our species (Francis, 2015). A descriptive analysis of the human body is enough to demonstrate the groundlessness of the idea of deficiency. I intend to show this by addressing four aspects: 1. the reorganisation of the muscular-skeletal set-up for bipedalism; 2. the transformation of the sexual and reproductive dimension; 3. the evolution of the neurocranium and the practognosis; 4. the development of the social dimension, in its affiliative, collaborative and mimetic aspects. But before examining these points, it is useful to reflect on the concept of specialisation in evolutionary theory. It is incorrect to assess the human species by separating it from its phyletic lineage, i.e. primates in general and *anthropomorphae* in particular. It is a *logical fallacy* to judge it through the lens of the tool dependencies it displays today, to conclude that it is a non-specialised entity. In order to understand the human degree of specialisation, it is necessary to objectively compare our morpho-functional structure with that of our closest relatives and to establish whether we have deviated from them and in what way. If we proceed along these lines, we will realise that human beings have reshaped their anatomy in many areas and have done so not in regressive-ancestral terms, but by acquiring specialisations consistent with a new lifestyle, as an adaptive function in relation to a new ecological niche.

This niche evolved over millions of years in relation to the occupation of a habitat, the savannah, that was radically different from the previous one (the forest). If we compare the human being with the chimpanzee, the bonobo, the orangutan and the gorilla, we will quickly realise that we are the species that has moved furthest away from the shared model – think only of our locomotion – and has done so by acquiring very precise specialisations. The human body is by no means larval, but was redefined on the basis of coherent performances and correlations, encompassing all systems, the functional cycles of metabolism and physiology, and ethological predispositions (Eibl-Eibesfeldt, 1989). It is incredible how people continue to talk about *non-specialisation* in the face of such blatant objective data! Indeed, it is incredible in some ways, but understandable from an ideological point of view. The deterministic conception of nature, which transforms the phylogenetic endowment into a limitation of the field of freedom available to the individual, suggests that humanity should be considered as poorly endowed with adaptive equipment, precisely in order to maintain as much space as possible for self-determination.

On the one hand, the mechanistic conception of nature is not questioned, because this would bring down the distinctive principle of the human – the anthropological machine (Agamben, 2004) – which needs the dark animal mirror to construct the edifice of *humana dignitas*, derived by opposition, disjunction and elevation. On the other hand, for the same reason, it is necessary to lighten human nature as far as possible, to the point of making it almost insubstantial, in order to sustain the free will and self-determination on which the humanist revolution is based (Braidotti, 2013). It is necessary to overcome both these positions in order to recognise that: i) the innate heritage does not diminish the subject's space of freedom, but on the contrary, makes it possible by supporting experiences both from a copulative point of view and in the form of *a-priori* content; (ii) it is thanks to the redundancy and specialisation of human nature that our species has been able to develop such a massive technical-technological dimension. The innate endowments represent the first equipment through which the individual faces the world, a toolbox capable of evolving, ushering in knowledge that is more adherent to the situation at hand. The innate does not limit expressive freedom, but rather makes it possible through the occasionality with which the individual uses phylogenetic endowments as evidence of adaptation to the here and now. The innate provides, on the one hand, the deductive hypothetical arsenal that makes it possible to deal experientially and not passively with the world, and on the other hand, the affective copulative dispositions that trigger experience through involvement.

So let us examine the four points above. The human species has innate characteristics – understood as morphological and physiological traits, behavioural tendencies, *Umwelt* dimensions, and a range of possible ontogenetic outcomes – just like all other species (von Uexküll, 1956). These are the result of the phylogenetic pathway, i.e. the selective pressures and particular lineage of derivation, of the species *Homo sapiens*. We must, therefore, speak of human nature as both the starting condition and the perimeter of the field of possible development for the human individual. The human condition is not fully liquid, but certainly plastic, and this *plasticity* can be traced back to somatic exuberance – for example the large number of neurons and the flexibility of the practognostic conformation – and not to some alleged incompleteness. This redundancy allows for a multiplicity of ontogenetic directions, i.e. individual and cultural variations, by virtue of the particular conditions that human beings may encounter in the course of their development. This plurality, however, is confined within a well-defined perimeter: we cannot transform ourselves into any entity whatsoever, but we can assume forms that, however plural, are compatible with a certain species-specific dimension.

From the point of view of *bipedalism* (1), a path of specialisation that began with the Australopithecines around 4 million years ago, we note a general remodelling of the postural and motor chassis that led to significant consequences from a skeletal point of view – the same ones that allow us to state that *Australopithecus afarensis* already had an erect position and moved in a bipedal manner (Johanson, 1981). Bipedalism, moreover, modifies the perceptual and operational set-up, for example by widening the field horizon and freeing the forelimbs from the locomotor task. This leads to a differentiation between the hindlimb train that specialises in bipedal posture and the forelimb train that strengthens its practognostic skills and improves eye-hand coordination. In fact, the entire anatomy undergoes a transformation, alongside the constitution of the foot, which is undeniably an extremely important specialised development, given the starting form present in anthropomorphs. Indeed, bipedalism triggered a series of transformations in a specialised sense that would differentiate our lineage from other anthropomorphs. Hence: i) a muscular-articular system that, starting from the heel, strengthens propulsive capacities and enables prolonged walking; (ii) the positioning of the occipital hole in the centre, allowing for standing; (iii) the transformation of the femur, which elongates with respect to the humerus, modifying the femoral tubercle so as to facilitate bipedalism even at speed; (iv) the transformation of the curvature of the spine with thoracic kyphosis and lumbar lordosis; (v) the remodelling of the entire pelvis and in particular the upward and backward expansion of wings of the ilium, enabling the anchoring of powerful gluteal muscles.

The sexual and reproductive dimension (2) also underwent an equally substantial transformation, aimed at reducing generalised sexual attraction and strengthening the couple bond. Here, too, we are faced with relevant transformations such as the neonatal degeneration of the vomero-nasal or Jakobson's organ, dedicated to intercepting pheromones and thus the readiness to mate, and the cryptic transformation of oestrus, which, being no longer manifest, lowers generalised sexual attraction. These are not negligible changes, because they indicate a transformation of the sexual engagement mechanisms of the species, in the specific direction of lowering recall. Conversely, the couple's sexual bond was strengthened through the activation of two promoting phases: a first dopaminergic effect that increases desire and euphoria; a second effect based on oxytocin that creates a condition of attachment. There are also further anatomical changes that many authors consider central in strengthening the couple bond (Morris, 1967), such as: i) the remodelling of the breast in an attractive function during coitus in the ventral position; ii) the enlargement and elongation of the penis; iii) the development of female orgasm; and iv) the loss of the penile bone.

These are not small changes, then. And there is more: it was above all the parental metamorphosis that sanctioned our unique specialisation, amplifying the predisposition to provide parental care, which is already emphatic in anthropomorphs. Our chimpanzee cousins give birth to fairly independent cubs with an encephalic volume that is about half that of adults, an acceptable compactness of the cranial bones and a level of myelination that enables them to cling to their mother and coordinate their first movements. The situation in our species is quite different. However, this is not a regression, but a specialisation that allows the human being to develop a considerable neurocranium while avoiding dystocic birth. Indeed, the remodelling of the pelvis due to bipedalism has defined a certain circumference of the birth canal, an insuperable constraint. The direction taken by the phylogeny of our ancestors was that of premature birth, quite similar to canids and felids (Portmann, 2022).

The human cub at birth has an encephalic volume that is about 20 percent of what it will be as an adult, its cranial bones are not welded together, as evidenced by its numerous fontanelles, and it has a neural connectome that is not yet myelinated, resulting in an absolute inability to coordinate motor skills. Our newborn would not only be incapable of clinging, it cannot even hold its own head up. Faced with such an immature infant, the entire ethological spectrum of parental behaviour must necessarily develop, on pain of extinction. It is essential to realise that the various body and behavioural traits are not disconnected from each other, but are interrelated and in a certain sense act reciprocally as selective pressure factors. An immature newborn requires a greater willingness to care for it (epimeletic motivation), this produces a greater sensitivity to paedomorphies and the latter, in turn, creates a selective pressure that rewards those paedomorphic cubs with a greater capacity to recall care (et-epimelesis). We must, therefore, agree with those philosophers who place the attitude of care as one of the fundamentals of human nature.

It is not easy to understand the reason for the development of the neurocranium (3) that now characterises our species, but which, in fact, testifies to a process that we find as a tendency throughout the evolutionary bush of the *Homininae*. In the comparison between human beings and chimpanzees, which as we know share as much as 98.8% of their gene bases, some authors have noted that in humans this functionality was reduced by a mutation in the MYH16 gene – encoding a particular myosin that plays a role in jaw strength in the temporal muscle. As we know, Evo-Devo theory considers development as a fundamental moment in the definition of the phenotype and, consequently, the establishment of a certain gene pool. Following this approach, we can deduce that a decrease in mandibular tension on the cranial bones during the embryo-foetal phase would have allowed the skull to become thinner and larger. In other words, there would have been a reduction of the splanchnocranium and a development of the neurocranium (Carroll, 2005). The emphatic growth of the neocortex and prefrontal lobe indicates the human redundancy in the connectome, where more than 80 billion neurons each make thousands of reciprocal connections, giving rise to a structure with an almost infinite variability of wiring. In fact, the great development of the entire neurocranium has led to a capacity that is three times greater than that of chimpanzees and gorillas.

It is not easy to identify the driving force behind this process and we are most likely mistaken when we look for causal linearity at all costs. It is more plausible to believe that a flywheel or autocatalytic effect is at work here, due to positive feedback. It is conceivable that a condition of increased environmental precariousness, especially in terms of foraging, has been a factor in selective pressure on the ability to devise new solutions to food problems. Perhaps a more open and predator-exposed environment, such as the savannah, also favoured the development of defensive capabilities. Clearly a greater cognitive capacity may have induced a greater ability to use tools, a practice already present in anthropomorphs. This habit may have ushered in new selective pressures on practognostic abilities, hand-eye coordination and, above all, abstraction: the ability to see the tool inside the stone. Technical specialisation is a very marked tendency in human beings, which is why it has often been suggested that the term *faber*, undoubtedly characteristic of our way of life, should be used instead of *sapiens*, which is a little too pretentious given the contradictory behaviour of our species. I would venture to argue, on the other hand, that the technopoietic tendency was the greatest driver of neuro-encephalic development, although I do not think it was the only factor.

Let us now consider the last point (4): the development of social and communication skills. The human being is certainly a technological animal and bipedalism is certainly distinctive of it, but I would not overlook another great property of our species, namely the collaborative and communicative tendency (Tomasello, 2009). The human being has developed empathic capacities and teamwork tendencies that are unparalleled in the primate world, so that a single human's adaptive and performative capacities can never be assigned to individual abilities alone. Like other primates – think chimpanzees and baboons – human communities present a complex organisation, made up of hierarchies, alliances, personalities and cultural traditions. In hominids, however, the affiliative structure includes an emphasis on cultural traditions and the role of elders within the community. In other words, the inter-individual relationship is much more complex from the point of view of the intergenerational passage of traditions also due to a greater availability of communication tools. The forelimbs, freed from locomotor tasks, not only become formidable operational tools, but allow gestural communication at a distance and haptic facilitation. The lowering of the larynx also enables humans to multiply their potential for vocal communication.

There is, moreover, a deep connection between the social dimension, the acquisition of cultural traditions, the breadth of communication tools, the development of empathic-immediative capacities and the mimetic tendencies of human beings. In order to understand this propensity – which has to do with creativity and will therefore be a central topic in this essay with respect to technopoietic imagination – it is first necessary to return to the parental specialisation illustrated above. A strengthening in terms of propensity and competence in caring activities always rests on a development of the parent's ability to fulfil the cub's needs and to put oneself in its shoes. The development of epimelesis, i.e. the dimension of parental care, entails a greater tendency towards identification. Now, if we observe a large part of cultural productions – such as dance, fashion, music, cosmetics, or rituals – we will note relevant borrowings from animal others that have constituted, so to speak, the models from which to draw the draft on which to work in a representational and interpretative sense. It is not a question of mere imitation of the performances and lifestyles of these animals, but rather of reworking and accommodating them in our own bodies and, more generally, in our own existential dimension (Marchesini, 2017).

We must, therefore, speak of a relational creativity that is based on identification, that is to say, on the empathic vocation that human beings have strengthened in part due to their parental specialisations and in part as a recursive effect of techno-poiesis itself, which made us accustomed to extending our bodies beyond epidermal boundaries. Mimesis is based on the psychotropic propensity for possession or projection in the face of natural phenomena, in a kind of trance or shamanic ecstasy. There is no doubt that this faculty represents one of the common characteristics of humanity, explaining many cultural, artistic and above all mystical manifestations and the sense of the sacred (Sax, 2021). The introjection of expressive styles borrowed from other species must be considered the first great flywheel of cultural evolution. Certainly, in order to make it feasible, it is necessary to possess certain predispositions towards identification, a strong mimetic motivation and good learning skills. On the other hand, somatic morphology can also favour this activity: for example, the aforementioned lowering of the larynx amplifies the number of vocalisations available, or the particular elasticity of the spinal column multiplies the possible kinaesthetic figurations.

Ultimately, we must admit that humans have their own somatic and cognitive condition that cannot be ignored or transcended, because it is the basis for the multiform expressions that characterise us. Human nature is far from incomplete, because it is the result of a long process of adaptive remodelling, as evidenced by the findings of palaeoanthropology. To consider the human species as the outcome of a history of specialisation is to deprive it of that liquid dimensionality that we find, for example, in the fanciful dissertations of transhumanism. Our way of interfacing with reality has a peculiarity that cannot be wiped away by the mere application of a tool. Picking up on the concept of *Umwelt*, we could say that the world, as it appears to us and as we conceive it, is not the expression of a neutral and universal character, but the emergent product of the organisation of a precise plan of reality. This can be transformed by processes of accommodation, for example through instruments that modify visual resolution such as the electron microscope, but it can never be completely revolutionised, because any technological graft must still remain within a range of possibilities. Greater restraint should therefore be exercised when making predictions about the reconfiguration of the human condition.

The aforementioned specialisations should not be forgotten. They speak of our precise place within the biosphere and, at the same time, of our dependence on existing ecological balances, which suggests that we should pay greater attention to environmental emergencies. On the other hand, it is precisely these species-specific dispositions – such as the plasticity of the connectome, neonatal immaturity, the social dimension – that have enabled humans to go beyond their natural heritage, i.e. to go beyond themselves, taking on new technomediated existential dimensions. Even the presence of certain motivations with a high copulative value – such as the tendency to imitate what they see in their environment, to adopt cubs of other species, to collect and catalogue what they encounter – have enabled them to have very close relationships with natural otherness, taking on certain styles from it and importing them into their own cultural heritage. These borrowings have found fertile ground in the human mind, becoming metastable entities capable of assuming their own evolutionary autonomy with respect to the demonstrator present in nature. What happens is an epiphany, where observing the flight of birds not only teaches man *how* to fly but, even before that, *that* one can fly, opening up a new metamorphic dimension to the human being.

One of the most relevant characteristics of the human condition is its desiring nature, a tension that nourishes ingenuity and the imagination and yet needs to be redefined with respect to its traditional reading. The desiring condition should not be understood as a lack that seeks a process of appropriation, but as a proactive and copulative propensity towards the world. In other words, the desiring condition is the wish to express oneself through deeds, to devote oneself to an activity, to involve oneself in action, to give oneself to active life. Unfortunately, a consumerist reading of desire has misled us into believing that desire finds fulfilment in appropriation when, on the contrary, it is nourished by dedication and donation. Human nature, therefore, has within itself the conditions to germinate the most edifying and satisfying works, but it can risk being led astray through deviations of its own tendencies.

The main mistake, in my opinion, is always to underestimate, in either sense, the potential of human nature, which is far from being characterised by ancestrality or larvality and the characteristics promoted by the humanist tradition. The perception of deficiency that leads us to believe that technology is compensatory for our incomplete nature is an interpretative bias: it means mistaking the consequence for the cause. It is techno-poiesis that produces *a perception of deficiency*, so technomediation should always be considered as exposure not as an exemption. There are many factors that produce such exposure – which can be framed as dependence on external support or a sense of need – some of which are performative, others related to social affiliation, and others still to ontogenetic integration. The fact remains that there is no point in talking about human incompleteness, either from a theoretical-evolutionary point of view or from a descriptive one. On the contrary, we must admit that it is precisely thanks to the specificities of our phylogenetic heritage that our great cultural and technoscientific achievements have been made possible.

CHAPTER 2

THE METAMORPHOSIS OF PERCEPTION

Premise

Every technological acquisition has consequences that affect people's entire experience, including their lifestyle, social participation, sense of security, satisfaction of needs, comfort levels, and leisure time. This creates genuine dependencies related to the perception of what is optimal. In practice, we become accustomed to performative standards related to the effectiveness or efficiency of practices, as well as the kind of achievable goals that are inextricably linked to technological support. As a result, all technology introduces a new *dimension of experience* to which we quickly become accustomed, considering it the new acceptable level of normality (Douglas, 2020). This creates a sense of need or incompleteness, making us feel naked or exposed if we are deprived of that technological mediation. However, analysing this feeling of lack is not always easy, as it is not solely related to the field of possibilities but primarily affects people's agency. In other words, it is a perception issue that extends across all of the subject's experiential zones, encompassing not only the available operational space but also the way in which they perceive the world and themselves within it, constructing a self-image and a vision of everything they intentionally relate to (Noble, 1997).

Of course, perception is not only the impression we have of things and ourselves – for instance, the accessibility of goals, the conformity of achievements, and the gradient of well-being to be expected – but also involves the existential dimension of the individual as a whole. Technology can either increase the depth of an experience, such as the horizon of communication offered by mass media, or it can distance us from a particular experiential field, such as the relationship with nature that was still present in rural society but has been lost today. The transformation of individual perception will also have an impact on affective and cognitive systems, such as the sentimental ties that experience creates or abolishes and the knowledge and skills it requires. Generally, every technological advancement alters the self's relational field, including the level of interaction with other individuals, the nature of social connection and belonging, the rites of passage that demarcate developmental stages, and the rituals and symbols of social encounters and status (Turkle, 2012).

Our social interface systems are linked to *proxemic space*, which refers to the level of interactive and operational proximity required by the individual to perform various social functions, such as meeting with friends, exchanging information, and feeling part of a group. Every technological change alters our interface systems. We saw this very clearly with the introduction of telephony. Landline devices had already distanced the space of interconnection, but the advent of cell phones widened this field even further, eliminating the need to stay in a specific location for a phone call. Smartphones have further amplified this interface through the use of social media and the ability to participate in events online. Thus, such innovations have not only altered the field of operation but also transformed the very concept of conviviality. When I speak of a metamorphosis of experience, I refer to a transformation that cannot be limited to the operational field, as it affects all aspects of the human condition. The introduction of any technology establishes a *new way of living* that is only achievable because of it, so it is accurate to say that every innovation generates a need and triggers a sense of lack.

2.1 People's experiential field

One of the most sensitive aspects of the introduction of new technologies is the experiential space, or the type of experiences that the technology enables, hinders, or alters. Experience is essential in constructing the Self, as it establishes the primary *vectors of individual learning*, including habituation and sensitisation thresholds, associative conditioning mechanisms, reward and punishment processes, factors that differentially influence predispositions, variations in affective dispositions, knowledge construction, and metacognitive development. Technology is always a *mediator of experienc*e, capable of modulating and sometimes altering the configuration of experience itself in terms of the type of experiential substratum, depth and gradient of immersion, scale of detail of reading, sensory systems involved, coherence with perceptual *Gestalts*, and operational affordances (Gibson, 1979). What Galileo's telescope and Antoni van Leeuwenhoek's microscope enabled should not be considered merely an enhancement of the eye, but a transformation of vision itself. In other words, technology brings a metamorphosis of experience that affects the interface between the subject and reality (De Kerckhove, 1991).

Another example of the transformation of the individual's relationship with the world through technomediation can be seen in the concept of *temporal experience*, which can be observed by comparing the same route travelled on foot, by bicycle, or by car. Each technology alters the *sense of time* and experiential detail, resulting in a profound impact on the individual's biographical structure. The rhythm of phenomenal alternation plays a crucial role in defining thresholds of arousal and attention-concentration, affecting not only the processes of acquisition and memorisation but also the pace of events to which one becomes accustomed. It is no coincidence that today's narrative plots in literature and film are noticeably more frenetic than those from thirty years ago, and watching a film from the 1980s may give the impression that events unfold slowly and predictably. However, this shift is not solely due to changes in cinematic techniques or narrative poetics, but is also a result of an anthropological shift in experience. The acceleration of technological performativity in the last century, particularly with the advent of digital technology, has introduced a sense of synchronicity in the unfolding of processes and has undoubtedly influenced this shift in storytelling.

The compression of lived time does not enhance the subject's experiences, as we might assume, but rather alters the *experiential framework* by emphasising certain aspects over others. For instance, a car trip emphasises the destination, while a walk brings attention to the journey itself. Furthermore, the emergence of the Web, with its capacity for rapid *information processing*, has fundamentally transformed the structure of memory, creating a sense of synchronicity between the past and the present. Many authors have described the phenomenon of a hyper-present or continuous present (Rushkoff, 2013), which has a constraining effect on people's lives due to the acceleration of decision-making processes, a loss of time for reflection, and an obsession with real-time updates on breaking news and polling data. This alteration of the meaning of the present moment results in a contraction of chronological space and gives experience a magmatic and precarious texture, accelerating our projection into the future. Whereas in the past, the projection of our journey was predominantly geographical in nature, today it takes place more in chronological terms, as if the cartography of reality had lost its consistency.

The transformation of our perception of time has altered the structure of our biographies, resulting in a more fluid experience that is less anchored in the present and more persistent or resonant. This is due to the multitude of digital traces scattered across various devices, which echo and return to the individual as if they were new. On one hand, the acceleration of the present has made our perception of time fleeting, while on the other hand, the echolalia produced by the countless self-images has made the past, which used to be experienced with little perception, the only tangible present (Floridi, 2014). The digital age has profoundly altered our experience of time, leading to a biographical structure that is non-overlapping between today's kids and the so-called baby boomers. After all, generations born in the second half of the 20th century have also adapted to a different time regime in recent decades compared to their youth. People over the age of 50 today have effectively experienced two distinct phases of life, divided by the new millennium.

Relationships and the social connective tissue have also changed. The emergence of social media has transformed how relationships are perceived, leading to a constant connection that erases physical distances and alters our very understanding of presence. This has resulted in a revolution in proxemics, making individuals more interactive but also increasingly enclosed in a somatic-narcissistic bubble that is ever wider and more impenetrable (Benasayag, 2015). The abundance of possible relationships has made them more fleeting and less solid, based on momentary connections and little solidarity, dissolving the Goethean idea of elective affinities. Conviviality has changed, with participation becoming more anonymous and unidirectional, resembling a buffet rather than true agapic participation. Presence has become a way to manifest the emphatic self through the spectacularisation of everyday life. This constant and generalised connection, akin to a trawl in the ocean of virtually possible relationships, has made the individual the only seemingly stable certainty, providing a foundation on which to seek meaning in one's existence.

We can no longer view technology as a humble servant that carries out our commands. Instead, we should acknowledge that it brings about significant shifts in our ontological definition and introduces new experiential perspectives. The narcissistic limitations that we observe in this era of fleeting digital relationships are a product of habits formed in adolescence, when the urge for social interaction is intense. In digital encounters, bodies no longer recognize each other through mutual acceptance of otherness; rather, images seek each other out to reflect *themselves*, reducing the person to its etymological meaning of mask and turning them into a logo to be marketed. The infosphere has inaugurated a new sacredness – an *ego-theistical* one – where the Self invents its own universe. Despite the undeniable resonance of Cartesian solipsism, the quest for online evidence to prove one's existence annihilates thought and doubt. All is nullified in the hypnotic reflection of a demiurge Self that has turned encounters into mirrors of its desires (Blackburn, 2014).

Technology is not merely a tool for achieving pre-existing, latent ambitions or desires that exist within human beings but operate under the radar. It is a phenomenological transformation of the subject through the alteration of their experiential field. It is through experience that the individual perceives themselves and takes on a *certain form*, thanks to acquired content. We could say that technology modifies the *Umwelt* of the subject, transforming not only their operational endowments (i.e., their agency) but also their immersion in the world context (Von Uexkull, 1956). The changes induced by technology do not simply juxtapose with human nature but imprint it with a certain variation, ushering in a new existential plane. As a result, with the emergence of this renewed manifestation of the self, bearing a specific attribution of meaning, it is no longer possible to return to the previous condition without feeling a lack of meaning or a sense of existential despondency. Technology changes the *Gestalt* patterns with which the individual faces their here and now, understood as a space of presence and participation. This implies changes in sensibility, agency, and relationality, but above all, new attributions of meaning (Lévy, 1990).

The attribution of meaning is always based on a phenomenological plane that depends on the individual's space of experience. Technology changes this interface, giving rise to a new phenomenological plane that transforms the entire positional spectrum of the individual. It's a mistake to view technology as simply an operational enhancement in instrumental terms. Experience is related not only to possibilities but also to the sense of rootedness an individual has built with space-time. For instance, how does someone feel when they don't have their smartphone with them? It's not just about losing the ability to communicate with others and a consequent feeling of helplessness (Cheever et al., 2014). They may experience a sense of freedom because they can't be reached, or a sense of precariousness because they're disconnected from their network of relationships. In either case, it's their experience that is liquefied, not just what they can or can't do. It's as though they are stranded on a desert island with endless time at their disposal and no chance to make use of it.

Examples abound of how technology has altered the perception and experiential interface of humans. Consider the mastery of fire, which transformed our ancestors' relationship with the world by expanding the size of the space they could inhabit and by providing protection against dangerous animals (Goudsblom, 1992). The use of chipped stone opened up new sources of food that were previously inaccessible, thereby changing how people perceived the act of foraging (Sahlins, 1972). Fire also had a transformative impact beyond its practical uses; by taking the role of a secure base, it radically altered the principles of habitation and was later personified in the deity Hestia. Even the domestication of dogs transformed the sense of space and perception of the world, giving rise to a new conception of residence and new standards for safety and environmental monitoring.

The metamorphosis of perception reflects a shift in the *experiential canon* that an individual can access. With the increasing sensory immersion provided by digital technology today – which risks becoming all-encompassing with the refinement of the Metaverse – it is clear that the structure of experience has undergone a huge transformation (Longo and Vaccaro, 2013). Analogue experiences, in fact, are very different from digital ones, not because they alter the content of music compilations, films, photographs, or books, but because they connect the subject with the medium. Taking a photograph using analogue methods involves coming into contact with the film, smelling the acids in the darkroom, working by hand, and acquiring technical skills. Listening to a vinyl record is not the same as clicking on a track; it involves a ritual that is accomplished through precise sensory and operational abilities – I still remember the care required to remove the record from the cover and to place the needle on it. A printed book can be smelled, stroked, and felt as one flips through its pages. Above all, it can be experienced in a syntonic way, because the paper book is not immaterial, but an entity that inhabits our physical world, resting on a bookshelf or speaking to us. Writing is also different when it is articulated through the movement of the hand on paper, and correction is never a simple tap on "delete", but rather a return to the word that leaves a trace.

To be clear, I am not being nostalgic for the past, playing the *laudator temporis acti*. I simply wish to emphasise that we are not only discussing performative differences, but a complete transformation of experience. This includes the way in which we live through and perceive our actions on the world and the feedback that this provides to our senses. What's changed is the dialectic of perception – the process that relies on neurobiological inputs that compare the feeling that something is out there with the effects of our own actions. The analogue constructs an experience that is indeed analogous, meaning it maintains some degree of similarity to what it represents. This is fundamentally different from the digital experience, which proposes a dematerialised alternative and seeks to assimilate the user into the Web.

Time moves at a different pace in the digital world, causing us to become accustomed to an overwhelming synchronous condition. The availability of digital technology freezes our presence, repurposes it, and *transforms our existence into resistance* (Marchesini, 2009). This paradox leads to the avatar taking precedence over the biological consistency of life. Today, people store an inordinate amount of photographs in the cloud or computer memory, far exceeding the possibility of viewing them all. Children born in this century have hours and hours of footage documenting almost every day of their existence. As a result, even memories of one's past take on a consistency that exceeds the possibility of being fully experienced. It would take another lifetime to trace all the scattered images of oneself, the hodgepodge of traces that persevere in their *resistance* on the Web.

Social media have revolutionised the very concept of contact and sharing: a post soon takes on a life of its own and navigates through time and into the windows of far-flung users, like self-replicating messages in a bottle. This no longer affects a small number of famous people or influencers, but is common practice: on average, people communicate with thousands of virtual interlocutors every day. Identity is no longer connected to its etymology, it cannot serve as a principle of individuation, but neither is it simulation, a mask or Batesian mimicry: it is simply something else (Gerbner, 1998). So can we talk about the technosphere only in performance terms? Evidently not.

Therefore, it is crucial to begin by considering the overarching implications of this metamorphosis, which not only impacts our perception of the world around us, but also our sense of self in relation to it. This includes our social relationships, experiences, the reconstruction of memories, and even our dreams. Technology alters our experiential dimension even before affecting our operational one, thus transforming the entire landscape before our eyes and giving rise to a new self (Kunstler, 1998). Perception means feeling oneself within a particular horizon of the world, where there is a close relationship between sensations and actions (Godfrey-Smith, 2020). At the neurobiological level, we already know that agency produces precise corrections in the perception system to avoid disruptions in orientation. This leads us to conclude that every performative shift is related to a change in perception.

Living within a recognised space allows individuals to define an accessible dimension, where they can have experiences, gather information, and, most importantly, encounter certain phenomena and not others. The experiential dimension is determined not only by sensory access, which is fundamental in establishing gateways to a particular way of living, but also by the knowledge and tools possessed by the individual. Therefore, the advent of new tools is bound to bring about a new way of being. Technomediation not only defines the realm of possibilities in terms of agency or what can be achieved, but also the space of experiential immersion and how we interact with others. The emergence of a new technology can reveal a new manifestation of the self, but this is only evident *after* the fact, and its effects on the individual's experiential and evolutionary level are unpredictable (Lifton, 1994).

2.2 The perception of needs

To explore the topic of the shift in self-perception caused by technological progress, it is crucial to examine the human needs that underlie it. To do so, we must consider the hierarchy of needs and how the sense of urgency or priority changes in different stages of life, in a dynamic perception logic that acknowledges objective demands, especially those related to somatic metabolism, as well as the subjective nature of needs. In fact, what is important to a child is clearly no longer important for an adolescent – just think of the all-consuming need to play and explore in childhood and, conversely, the irresistible yearning for affirmation that emerges during puberty. Similarly, priorities take on varying connotations throughout life, until they assume an entirely different profile in old age. The evolution of needs as we grow older does not just follow a linear pattern of stages that repeat generation after generation, but adapts to changes in the context (Piaget, 1970).

Our environment is also impacted by technological advancements that continuously emerge and alter the appearance of our surroundings and the relationships we have with them. These changes affect our perceptions, practices, and often our priorities, since many of our needs are driven or sustained by practices that may become obsolete when a new technology supersedes them. As a result, the projection, or the way we feel about a need, changes even if the underlying need remains the same. For instance, consider the search for coins and a free phone booth in order to make a call prior to the widespread adoption of cell phones. I am referring to the ritual significance of phone-booth communication, which was prepared through a set of specific operations: waiting for one's turn, preparing oneself psychologically, and marking the time of the conversation with the cost of the phone call. These projections and practices were suddenly lost with the advent of cell phones. Of course, communication is now facilitated, but its overall meaning has undoubtedly changed, and the perception of the need for communication itself has undergone a drastic transformation.

Therefore, we must speak of a *phenomenology of need* that cannot be reduced to a mere demand for a performance that facilitates or enhances the attainment of a goal. Rather, it concerns the way in which the subject's projection appears in their experience. Need is never an objective lack that presents itself as an attractor, but rather a tension that the subject produces by clinging to the scaffolding of their perceptual apparatus (Merleau-Ponty, 2003). Need is the way in which individuals recognize themselves in the set of relationships they engage with in the social and environmental fabric. Technology does not respond to a need but modifies the connective organisation of the self and, consequently, the coordinates of projection. At times, this metamorphosis annuls certain tensions that previously held priority value for the individual, because they stood at the crossroads of fundamental needs. At other times, it ushers in perspectives that were previously completely obscured by the narrowness of the landscape, but then suddenly become alluring or inescapable. In all situations, we certainly observe a modification of the *existential connectome of the person*, which radically transforms the horizon of possibilities and agency with immediate availability (Simondon, 2020).

In order to understand the anthropological changes induced by the technical evolution of societies and, consequently, to develop a philosophy of technology that goes beyond the compensatory Prometheanism typical of humanism, it is essential not only to question the theory of the incompleteness of human nature but also the explanation of projection as lack (Gehlen, 1988). This view leads one to see need as something that precedes the advent of technology when, on the contrary, need is a *product* of technology.

There is no doubt that human beings have certain needs that are essential for survival, such as the need for food, and others that are expressive, such as the need for exploration, play, or physical activity. However, we must acknowledge that it is often difficult to draw clear distinctions between the category of vital needs and that of mere expectations, not to mention the territories of desire and purpose. Often, these categories seem to blur into each other, or even belong to a different category depending on how one interprets them. Our projections are related to our sense of security, satisfaction, attribution of meaning, and the fullness of existence. Even though they may not involve basic functions such as eating, drinking, and sleeping, they can still be a source of malaise and discomfort. In turn, this can cause psychological problems such as depression or anxiety, which inevitably affect the whole organism.

Reflecting on needs is fundamental to understand how a person's perception of well-being evolves. In 1943, Abraham Maslow elaborated a psychological theory based on a hierarchy of human needs, using the image of a pyramid. At the base, he placed the fulfilment of physiological urgencies, followed by the need to feel secure, both physically and psychologically. At the third rung, he placed the sphere of belonging and the need for intimate relationships and friendship. The fourth level included the need to have a good degree of self-esteem and to enjoy the respect of others. Finally, at the top rung, we find the aspiration to self-realisation.

It is certainly easy to distinguish strictly physiological needs, which are imperatively imposed on the individual, from those needs that are more related to the life project, such as expectations for the future, desirable goals, objectives, and the purposes that give meaning to one's existence (Frankl, 1984). The latter type of need is more susceptible to cultural and social variables, which affect the matrix of relationships on which the individual relies to pursue various aspirations. The pyramid proposed by Maslow organises the first four levels as deficit needs, which are linked to objective deficiencies that must be compensated for, while the fifth level is associated with aspirations that arise from the individual's free proactivity. Maslow's intention was to identify a hierarchy of objective deficiencies, where higher needs can only be fulfilled when lower ones are met.

However, if we accept this model of needs, it would follow that all human actions, including imagination and technical creativity, are driven by the need to compensate for objective deficits. This would lead us to the conclusion that technological progress is the result of partial solutions rather than a pursuit of greater possibilities. This reasoning is clearly flawed, and we must acknowledge its basic fallacy: the compensatory idea. Moreover, in addition to the supposed objectivity of needs, Maslow's hypothesis contains two other controversial points. Firstly, the hierarchical conception of needs fails to account for the different moments of life that disrupt any static-pyramidal arrangement. Secondly, the idea of self-realisation is over-rationalised and seeks to subsume the various passions that support an individual's life project (Marchesini and Celentano, 2021).

So let's analyse these two points, beginning with the hierarchical conception, and examine whether it is applicable to the behaviour of other species. Ethology refutes this hierarchical linearity, showing that the urgency of a need does not depend on the satisfaction of previous needs, as would be inferred from the pyramidal structure. Instead, it depends on the strength of a certain disposition in the economy of the moment or in a given expressive phase. For example, during the courtship period, many species tend to prioritise mating behaviours and put foraging and defensive behaviours on the back burner (Tinbergen, 1965). Among animals that live in large communities, social affirmation may override the mechanisms of physical need and focus on survival. Animal behaviour does not follow a consequential linearity in setting priorities but rather responds to physiological predisposition mechanisms regulated by endocrine and neuromodulator systems. Some hormones, such as testosterone and vasopressin, increase competitive tendencies and decrease the sense of hunger. Even oxytocin, a neuromodulator often trivialised as the "love hormone," exceptionally increases the sense of security, leading individuals to put their lives on the line to defend cubs or other pack members (Carlson, 2013).

Maslow's theory associates need with lack, in which the most critical needs, corresponding to the first four levels, are defined as gaps that need to be filled (Broom and Johnson, 1993). A different approach, which does not rely on a hierarchical structure, is the concept of welfare introduced in the 1965 Brambell Report, which is based on five freedoms: 1) freedom from hunger and thirst; 2) freedom to have an adequate physical environment; 3) freedom from pain and disease; 4) freedom from fear and discomfort; and 5) freedom to manifest the species repertoire (Brambell, 1965). This criterion was designed to establish a welfare canon for application on farms and comes closest to providing a comprehensive view of needs (Maslow, 1954).

The second criticism of Maslow's theory is related to self-realisation, which Maslow places at the top of the pyramid as a form of self-affirmation. However, this is not the same as the *species-specific expressive need*, which is a much more complex and nuanced urge to manifest one's innate characteristics (Weaver, 2016). The expressive need is not just about self-realisation, but about the opportunity to fully express one's nature, which is essential for well-being. Maslow's pyramid, on the other hand, despite its focus on intrinsic needs, is focused on compensating for deficits and is far removed from the concept of expressive need.

To avoid the absurd separation of human beings from other animals that are taxonomically similar, it is clear that the need for expression arises from the need to carry out basic actions that are specific to each species (Lichtenberg, 1989). For example, a cat's predatory behaviour may be expressed in hunting or playful behaviour toward a spool or ball, but the driving force behind it is the pleasure of chasing, not the lack of prey. While in some situations, the need for food and the pleasure of chasing converge in the feline's hunting environment, in other situations, the two needs do not overlap. Thus, even with a full bowl, the cat may still experience a predatory appetite, meaning it will have the desire to chase.

The desiring copula, therefore, does not arise from a specific deficiency or attraction exerted by the goal, but rather from a projection of the individual supported by a verbal predicate – in the example above, the desire to chase – from which proactive expression arises. The same is true for humans. We have intrinsic tendencies or motivations – such as collecting, caring, competing, and cooperating – which, like verbal predicates, underlie a wide range of actions, whether they are goal-oriented or simply playful.

When we observe children's behaviour, we can notice that, despite their almost infinite *modal variations*, they always rely on the same verbal predicates to support their actions. For example, they might collect different things in various places using different modes and skills, but they are always *collecting* something. While the verbal aspect remains constant, the conjugation terms vary depending on the person, historical period, culture, experiences, and social or generational group to which they belong.

Thus, expressive needs are not necessarily related to self-realisation, but they are certainly among the strongest passions and yearnings of the individual (Keverne et al., 2005). These needs have a dual nature, combining innate motivation with learned expression, which makes them difficult to define but nonetheless crucial in providing meaning to our lives. We all find satisfaction in our passions, and even more importantly, we find an existential centre of gravity in the act of desiring. In fact, it is not just about acquiring something, but rather about a deep dedication to an activity that fulfils us at a fundamental level (Marchesini, 2018).

Technologies do not alter the innate verbal predicates of human beings, nor can they do so – let me reiterate this for those who believe that our desires solely stem from rationality and elaboration. However, technologies can significantly impact how these desires are expressed, by shaping goals and situational complements (Benasayag and Schmit, 2003). These devices do not affect the fundamental predicative dimension of humanity, but rather alter the type of projection by manipulating expressive factors such as goals, the structure of practices, gratifying contents, and the perception of optimal performance in terms of efficiency and effectiveness. In other words, it is the connective structure binding individuals to the world that is modified. Consequently, the perception of the self is transformed.

Technological development cannot satisfy or calm, let alone nullify, the projective impetus, but only shift the trajectory of projections. Shifting projections to more distant territories means giving rise to an increasing sense of precariousness, because it moves us away from the phylogenetic centre of gravity or, if you will, from the body's potential. This is to reiterate that technomediation has never been a crutch to make up for the shortcomings of a deficient nature; rather, it has always played the role of a driver of needs that increase the individual's exposure to external influences. The outcome of technomediation is not a dressing, but a stripping, amplifying dependence rather than autonomy (Sloterdijk, 2011).

Furthermore, this is not a negative consequence if one is fully aware of the dependence that is being contracted. In my opinion, fragility, the need to establish relationships, and connection to the world represent fundamental parts of being human. However, it is a mistake or, better, a blunder to view the technosphere as a protective amnion, a disconnection from the biosphere, an armour that shields humans from external contamination, or an alchemical process aimed at extracting human essence. This viewpoint is clearly flawed.

Technology increases exposure, which in turn increases the sense of lack, causing a transformation in our perception of needs. This leads us to imagine a horizontal trajectory of the human being in the world, rather than a verticalisation above it. However, if this openness to the world is not supported by awareness and is misunderstood in a solipsistic sense through the belief in compensation, the illusion of technology as a saviour that can bring about self-realisation becomes a dangerous source of suffering. The reason for this is simple: it induces narcissism and a tendency to appropriate things (Lasch, 1979).

Indeed, the explanation based on lack is not only wrong, but it is also misleading and can cause great sorrow. In fact, it conditions us to believe that happiness comes from the objects we own, the accomplishments we make, and the functional devices we need to keep up with our projections. The consequence is far from unproblematic: once we attain our goal, especially if it is easily achievable, we will feel dissatisfied and more vulnerable. The reason for this disappointment can be traced back to the fact that we base our desires on lack, that is, on the object or tool that allows us to achieve a certain result (Gilbert, 2017).

On the contrary, the real focus of desire is *action* linked to predicative expression: the only thing that can calm restlessness is to engage in the activity itself. Tools, therefore, risk feeding the sense of dissatisfaction precisely because they lighten or limit the expressive load, making things effortlessly accessible. Why is it commonly said that difficult challenges have a higher value? Certainly not because the goal itself is inherently more worthy, but rather because satisfaction comes from action and desire is fuelled by projection, not attraction. The more tools are used to diminish the expressive range of the subject, the stronger will be the sense of dissatisfaction. That, in turn, will lead to raising the bar of desirable goals, which are inevitably destined to disappoint us once more.

A similar argument can be made for emotions and feelings, which, more than any other psychological disposition, define the subject's state at a certain time or in a given circumstance. Again, it is important to distinguish the affective disposition from its expression: it is one thing to be afraid; it is another to face what causes this feeling and how we express it. Such a distinction shows how we can never experience a complete sense of security or happiness, because fear is part of us and we cannot eliminate it, just as joy cannot last indefinitely and cannot be reliably caused on command.

The affective component, in fact, depends only to a small extent on the circumstances that arouse it. Mostly, it is regulated by the body's physiological systems and depends on one's character (LeDoux, 1996). In other words, we do not learn fear, but when and what to be afraid of. No technological protection is able to extinguish fear: what it can do is modify the expression of this emotion. What changes is always the projection, sometimes even by making up reasons or feeding ghosts in order to justify a feeling. Except for cases of dysthymia, the life of a human being cannot permanently dwell in a given emotional state, because the affective condition is a continuous fluctuation of feelings.

So, technology frees us from some expressions of fear, but not from fear itself. Through its protective devices, in fact, it only projects the feeling of precariousness onto other possible conditions of risk, often ushered in by the transformation of perception produced by *technomediation* itself. Sometimes this process of translating the phobogenic factor does anything *but* lighten the burden of fear, for nothing fuels terror more than the lack of a clear and explicit representation of what frightens us. This is why children need fairy tales dotted with monsters to tame their fears.

Both Maslow's theory and the Brambell Report refer to a generic liberation from fear – a laudable purpose if referring to life under constant threat, but a chimerical New Age claim if taken literally. In fact, losing the sense of fear is one of the worst alterations that can occur as a result of a neurobiological disorder (Damasio, 1994). The development of new devices does not eliminate fear but exacerbates it, and the reason is simple: by raising the level and number of functions required of the individual, it increases their sense of insecurity and ineffectiveness. Social competition is not mitigated by technological development, but taken to increasingly higher levels of confrontation, placing the individual in a kind of permanent balancing act, which is anything but soothing.

The perception of need is directly related to one's parameters of ability, that is, those basic skills that enable one to bear the brunt of the functional demands of society. Today older people struggle to navigate the mechanisms and procedures required to perform even the most common activities, which conversely do not frighten younger people (Slettemeas, 2014). The transformations of technomediated practices are easily assimilated up to a certain age, then this pliability fades away and people tend to live in the past, narrowing their field of operation and regretting a world that no longer exists.

Technology therefore creates a new need: to keep up to date, staying abreast of ongoing technological transformations. Until the end of the 20th century, this process followed a largely linear evolution, then it suddenly began to accelerate, generating a need for constant updating (Rosa, 2010). Clearly, the technological equipment of a society has a direct influence on expectations, demands, and habits that are taken for granted. Technology not only solves problems for us, creating a sense of dependence, but it also creates new problems. For instance, it often compels us to learn new procedures and adapt our work or daily lives to practices that further reinforce our dependencies.

2.3 Body perception and technology

In the following chapters I will analyse the phenomenon of techno-addiction, trying to avoid both the interpretive simplification of enhancement and the opposite sentiment due to the process of somatic down-regulation. Evidently, technology amplifies operational and immersive access, leading to addiction (Small, 2009), and it is equally clear that the externalisation of a function, from the body to the device, lowers the somatic exercise of that performance by down-regulating it. In this section, however, I would like to focus on another issue, which we can call propaedeutic to those topics: I am referring to the metamorphosis of body perception as a result of technomediation.

To us, the body is not an objective entity, but rather a construct resulting from the relationships we establish with external reality. There is the body as more or less accessible to the gaze of others, and there is the body that we perceive as a dimension of existence. Technology modifies this perception by altering the coordinates of the somatisation of experience: immersion in the context-world, sensitivity to phenomena and things out there, agency, and the gradient of proximity, relevance, connection, involvement, projection, and completion with respect to events in the temporal flow. We can say that each device produces a *Gestalt* revolution in our *Dasein*.

The invasive nature of technology with regards to the body leads to a shift in ownership and perception of the body itself, as the hybridisation of flesh and instrument alters our bodily experience. Once we have incorporated technology into our bodies, we cannot return to the previous state of being, but only continue on this new path, as the body has lost its previous cohesion, evolving to function on a hetero-referential matrix (Broadhurst and Machon, 2012). This *sense of loss* is felt when we recognize the organic insufficiency of our bodies without the device, and the complementary need for it. Removing the device would thus be like an amputation, causing a phantom limb effect. We would not be reclaiming our primal body, but rather realise that we have lost it, at least in part.

The implementation of technomediation inevitably leads to a shift in somatic perception. This can take two forms: i) a rise in the standards of performance, where the individual becomes accustomed to a certain level of access to the world and internalises it as the new norm; ii) a change in social connection, where the affective-relational texture is altered, and the individual becomes more reliant on devices that can ensure that level of participation. The introduction of a device always has this two-way effect, as any performance has an impact on the individual's sense of belonging to the social fabric, just as any relationship is evaluated based on the person's well-being and overall sense of conformity. In other words, *technology changes the perception of the relationship between the body and the world by altering the points of intersection.*

The transformation of perception brought about by technomediation creates an ontological illusion that can lead us to see it as a means of completing the subject and compensating for its deficiencies, thus enhancing the individual's power over external reality. However, this seemingly commonsensical interpretation is prone to multiple distortions. The most common error, as I mentioned, is explanatory inversion, mistaking the somatic plane of perception – a consequence of technomediation – for the cause that prompted its implementation or marked its success. Such an approach sees the technological or cultural device as a crutch meant to fix an *a-priori* deficiency, leading to explanations based on a compensatory or exonerative conception of *techne* (Lombo and Giménez-Amaya, 2014).

The second error concerns the way in which the hybridisation between the body and technology is conceptualised, as if it were merely an act of filling, i.e., the insertion of content into an empty container. This bias leads to the mistaken principle of inverse proportionality between container emptiness and the possibility of content introduction. This explanation is related to the image of a vessel to be filled, where empty space is what enables acquisition. However, this model has been superseded by the *principle of virtuality in complex systems*, where it is precisely redundancy that allows for the possibility of expression in a form.

The third error, which is also the most serious, is based on the concept of technological enhancement of the human body. There is no denying that technology can improve certain operational and interface performances, producing extensions of agency or a higher level of precision. However, it is how the concept of enhancement is viewed that can lead to misunderstandings. Enhancement is often considered as tantamount to: i) reducing exposure to external variables, ii) lowering the level of need, iii) providing greater protection from disturbing factors, iv) limiting functional variation, which is typical of the animal world, v) increasing decontamination from extraneous elements with a consequent increase in the purity and authenticity of humanity, and vi) increasing disjunction from the biosphere.

This is the model that has been transmitted to us by the humanist vision, where the concept of *dignitas* is accompanied by a series of qualities underlying ontological anthropocentrism (Debes, 2017). In this sense, reducing exposure and need, protecting human beings through a technological exoskeleton, and differentiating them from other animals are all consistent with the Promethean assumption. However, this is a grave mistake because technomediation, even in terms of extending access, produces the opposite effects, *increasing* needs and exposure, decreasing protection and disjunction, and contaminating the body.

Indeed, technomediation alters the body's extension and functional organisation, so autonomy is not increased but rather diminished, leading to *heteronomy*. The body incorporates the external element, becoming dependent on its contribution in the evolved conformation, much like a fish's gills entail dissolved oxygen in the water and make the fish dependent on this characteristic of the world. As the subject becomes more connected and exposed to the world through technology, they also become more needy and experience a transformation of the body, resulting in a reduction of ontological autonomy.

Technological development makes our bodies increasingly reliant on external formatting, much like a text that cannot be understood without a comprehensive set of notes. This is the hybridising – and not simply enhancing – effect of technomediation. Heteronomy characterises the phenomenology of life, as each organism, through adaptive specialisation, becomes reliant on the external context, already presupposing specific inputs from the world (Gilbert and Epel, 2015). Technomediation merely reenacts this process of dependence on external contributions to increase ontological space.

It is essential to consider that existence is not limited to a fixed perimeter but is characterised by ontopoiesis, where *Dasein* constantly transcends the present condition. The somatic aspect of this process is not merely a physical occupation of space but rather the emergence of a space of projection. The subject's movement is always the result of an *implicit peripatesis*, a tendency to project oneself towards conditions that have not yet been experienced. This disposition towards projection is not inherent but rather emerges from the human condition. I am referring to the psychological mobility of the subject, which causes their restlessness and curiosity and drives them to go beyond themselves (Siegel, 1999). This means that technology alters the peripatetic coordinates of the subject and, consequently, their *Dasein*. The body is a dwelling not only because we inhabit it, but also because its extension defines the habitable space of the subject. The moment we alter the coordinates of the body's extension, which are never limited to the epidermal perimeter or physiological centre of gravity, the subject's projection space inevitably changes.

Technology not only extends habitable space but also modifies it. The principle of *peripatesis* thus undergoes a most significant mutation, because projecting oneself into a space not only provides access to new configurations of reality but also results in reflecting oneself differently in this renewed world. Perceptual metamorphosis, therefore, concerns each individual's image of their body in space, with regard to extension, proxemics, sensitivity, and agency (Capucci, 1994). If I were to describe this transformation, I would define it as an experience that initially has an ecstatic quality: it's as if the sudden transformation of space jolted us out of our bodies. Later, through the introjection or somatisation of the device, we experience a feeling of possession, as if the device invaded us and then reconnected us to the common sense of extended habitation. Figuratively speaking, it is as if our flesh migrated along new textile weaves: in this metamorphic process, the boundary between body and device gradually disappears.

There is always an ambivalence between possessing technology and being possessed by it, but this is an epistemological illusion. Integration is not a completion or juxtaposition, but a mutual dimensioning. It is important to reiterate that somatisation is not just the phenomenon of filling the body's porosity, but a physiological interpretation of the device. Perception involves all somatic apparatuses, which are intimately related to each other. Therefore, we should not think of somatisation as only a psychic condition, as the endocrine and metabolic system are greatly affected by this physiological transformation. The device is not introjected neutrally and non-specifically but according to the specific dictates of that particular organism. In other words, the relationship between the organism and the device is reciprocal, much like the mirroring we observe in the structuring of a tree canopy: the oak tree transforms light into polylobate leaves, and the latter speak to us of a very specific kind of light. To understand what a technology can do, we must ask what a body can do (Deleuze, 2010).

As each invention changes somatic ownership, it is clear that the sensation of impropriety can never be an *a-priori* datum, which would conflict with the Darwinian conception of qualities, but rather a perception that arises from the extension itself. We have seen how it is quite bizarre to call the structure of the human body deficient or unspecialised, and it is equally absurd to compare it to a contralateral entity, animality, which is a kind of *superanimal* that subsumes the qualities of different species, serving as an unlikely chimaera for tautological comparison (Derrida, 2008). The feeling of deficiency does not arise from the way human beings mirror themselves before the archetypal animal, but takes shape in the face of the mass of functional exposures that the technosphere produces. It is important to avoid the bias that confuses *post-hoc* for *propter-hoc,* mistaking the consequence of technomediation for the cause of our need for the instrument. In my opinion, however, even worse than the epistemological mechanism of explanation is the idea that the device lowers our neediness or exempts us from something. This is an erroneous conclusion that arises precisely because of the compensatory meaning that is often assigned to the device. In reality, the device pluralistically introduces new needs by changing the perception of the self and thus the coordinates of somatic dependence on the world (Braidotti, 2013).

The advent of a device always brings about a phenomenological space that is inaccessible to the body without its support, and this initiates a transformation in numerous fields of perception. It is the landscape that stands before us that changes, as well as our relationship with the world and consequently with ourselves. For example, the telescope is not only an enhancement of the eye, but implies an epistemological revolution that takes us beyond the geocentric dream (Wolpert, 2000). Similarly, the habit of dressing makes us feel naked when we undress, inventing the very concept of nakedness. This sensation is not only about the ability to endure the rigours of winter and make possible the colonisation of otherwise inaccessible geographic areas. It also brings out new dimensions of identity, such as a sense of belonging, modesty, the nomadic perspective, and the alliance with animal deities, to name a few. On the other hand, the consequences also recursively reinforce the centrifugal process, as having clothes allows one to move even further away from the constraints of the phylogenetic habitat. Therefore, we must consider this hybridisation-induced perceptual transfiguration as a flywheel that amplifies the centrifugal process. The shift in perception becomes an incentive to unmoor and accept the challenge of pushing out into the open sea, to set one's existence as a metamorphosis of the flesh.

We do not only experience a localised mutation related to the specific body-*techne* relationship, but rather a systemic transformation of the overall perceived body image. Furthermore, this change is reinforced by a flywheel effect, where each technomediation paves the way for the next. It is not the insufficiency of the body that drives technological development – otherwise, we would observe a decelerative effect over time – but rather the somatic centrifugation that occurs as a result of technomediation, which in turn facilitates new technopoietic processes. To view the emergence of technology solely in terms of the performative qualities it allows for – and often imposes – is to oversimplify the evolution of tools. Such a perspective obscures the complexity of their development. We tend to connect the creation of a tool with the function it later serves in the technomediation process, presupposing a prescient imagination that had already predetermined its role. However, in reality, techniques that originated in one context often transform the technomediation space in unexpected ways, bringing out entirely new dimensions.

The epiphenomenon plays a crucial role in the metamorphosis of perception brought about by technological advancements. As often seen in the recent history of computing, inventions always come with collateral effects, not just exaptation phenomena (Gould and Vrba, 1982). However, the epiphenomenal dimension should not be trivialised, as these unexpected collaterals often prove to be the most productive areas in inducing subsequent technopoietic processes (Amzallag, 2021). This realisation highlights the *coevolutionary* relationship between human beings and technology, as each innovation brings forth new experiences, altering our perception of reality and consequently, our entire matrix of connection and projection (Nuismer, 2017). Believing that technology is solely about extension or enhancement leads to incorrect evaluative models of human nature, based on the assumption of empty inclinations and misleading perspectives for the future.

The perception of deficiency implants the idea of an unrealised human dimension, a destiny that must be fulfilled through the ever-increasing enclosure of the human being within a technological exoskeleton, to the point of merging with it in a post-organic conclusion that will finally distinguish us from other animals in every respect. To address the more complex phenomena that create a genuine need for the body to receive help from the device, we must understand the inconsistency of this line of reasoning, which aims to achieve ontological autarchy while increasing dependence on the outside.

2.4 The metamorphosis of the relational interface

We usually think that our sensory organs are the only interfaces that connect us to the world, but this idea is quite reductive: indeed, we see more than what is visible to the eye (Merleau-Ponty, 1964). Besides, the opposite is also true, as sometimes we even fail to perceive something that is right in front of us. *Gestalt* psychology has illustrated this concept through numerous examples of optical illusions and amodal completions, showing that seeing involves assigning a shape to a conglomeration of referents, rather than analysing each stimulus individually (Koffka, 1935). These perceived forms, which are attributed based on structure and significance, facilitate recognition, representation, and orientation, and are therefore valuable for mapping. They often evoke emotional and motivational responses, and thus elicit behavioural responses. This suggests that *Gestalt* processing carries *a-priori* value (Tinbergen, 1965).

Sensory and cognitive media cannot be divorced from the underlying message (McLuhan, 1964). This means that each device, by intervening in the interface media, transforms the intentional structure of the subject, that is, the content to which it refers. This awareness disrupts the merely operational conception of technology, which holds that every innovation simply amplifies the domain of what can be done without affecting the subject. This is the trivialisation of *techne* based on the humanistic paradigm, which assumes the ontological impermeability of humans.

Instead, perception is not a mere collection of objective data, but rather the creation of a *phenomenological layer of reality* that is imbued with meaning by the observer. It is not a passive exposure to the world, but rather an active *election of the world* (Lotto, 2017). This is not to suggest an idealistic notion of creative solipsism, but rather to emphasise the ability of the individual to bring out a version of the world, through their interface apparatus, that is extracted from the virtuality of reality. In other words, reality can take on various configurations, and what is perceived depends on the phenomenological reality constructed in relation to the interfaces that are utilised. The virtuality of the real does not imply an undefined condition that is completely open to the subject's configuring action, but rather a *range of potential configurations* (Deutsch, 1997) Perception, therefore, is the selection of one among the various possibilities that reality offers. It is neither an inductive exposure nor an arbitrary invention.

Thus, perception is the attribution of form and meaning that occurs through the sensory and cognitive systems of the subject, whereby they respectively access and process reality. Technology modifies the interface between the subject and the world, thus affecting both access to, and the *Gestalt* processing of, reality. This dialogue between the subject and the world, as said, takes place on a precise phenomenological plane, based on an ontopoietic impulse of the subject (Heil, 2003). The relationship that the individual inaugurates when perceiving reality is never limited to the simple ascertainment of what is present and manifest. Instead, it becomes *an active search for what could be* (Panksepp and Biven, 2012). The perceptual effort is always ontopoietic, projecting toward change to construct new existential dimensions. Perception is not merely a static monitoring of what is there, but a movement of the subject rummaging through appearance to bring out possibilities. We must not overlook the affective aspect of feeling and desire, the impulse for renewal that sustains the very act of facing the real (Gibbs Jr., 2006).

Technology brings about a metamorphosis of this space of motion or projection by changing the phenomenological plane of reality, thereby creating new possibilities for existential mobility that transcend the subject, *going beyond the already given* (Junger, 1932) This is achieved through the construction of new scaffolding in the world, where the external element brought by the technological hybridation process enlarges one's horizon and field of projective possibilities. Technomediation transforms the subject-world dialogue and constitutes a new plane of reality for exercising one's existential nomadism. It can be argued that in this process, the subject follows a thread of progressive decentralising, gravitating less and less around the phylogenetic barycentre (Anders, 1980). As this centrifugal process unfolds, the need for external contributions to realise the subject's existential plan increases. Contrary to the belief that technology increases the autonomy of the human being, it actually works in the opposite direction.

And here, in my view, lies the greatest error of the disjunctive view that characterises the humanist explanation of humanity’s technical-technological evolution. The suppletive conception of technomediation, as seen, begins with the hypothesis of our species’ deficiency, relying on Prometheus for ontic compensation and seeing technology as a crutch (Scheler, 1970). However, this approach inevitably leads to the ascensional presumption of humanity and the arrogance of divergence, viewing technology as a means for asserting our dominance over nature and definitively creating a gap between us and the latter, in line with Francis Bacon. On the other hand, if we view technological evolution as one of the *peripatetic expressions* of humanity, constantly striving to surpass itself through the introjection of otherness, as in other cultural practices, we can realise that technomediation actually enhances the interweaving of our relations.

Technomediation is thus a hybridising process that opens up new existential dimensions, which are always emergent and allow access to unpredictable spaces of existential nomadism. It's evident that this process, which is *not* compensatory, increases the relational field of the subject and their dependence on external factors. Therefore, it's a mistake to connect the ontological expansion of the human being made possible by technomediation with increased autonomy and self-determination. This is because technological development, on the contrary, makes us *more* reliant on external input to realise the new ontological plane and attribution of meaning we have acquired and are unwilling to let go of. The ontological plane is, therefore, co-constructed by leaning on the technological referent, so that we can view our existential landscape like a funambulist moving on a tightrope. Our hybrid condition makes us increasingly open to external contribution, more connected and anastomosed to the world, and more inclined to experience the beyond-human (Nietzsche, 1917).

From this perspective, we realise that the subject is fundamentally defined by relationships. Hence the need for an *eco-ontological* approach that reexamines the dimension of technomediation (Marchesini, 2021). As *techne* transforms the entire interface process, we can no longer speak merely of operational empowerment. We exist in the technosphere like spiders in their webs, where the latter (like the former) provide not only sustenance but also outline the living space, perspectives, pathways of movement, sensors about what happens, and structures of connection. The web is the spider's ecology, and every variation in its form is a metamorphosis of the spider itself. Spiders do not adapt to the world through their webs, but rather adapt the world *to* their webs. Lived space is not a sensory perception of the environment, but a projection onto what is accessible. Technology undoubtedly alters this accessibility. Each introduction of a new technology, albeit to varying degrees, always transforms our projection and yet leads us to believe that we have simply increased our capacity to investigate reality.

Let us consider an alternative hypothesis, whereby what we define as real is, in fact, a *highly virtual* entity that can take on multiple configurations within a limited scope. Each variation would be determined by the investigative approach adopted, defining an image or plane of reality that, as mentioned earlier, depends on the interface. Altering the mode of connection with the world, such as through new technological devices, would not reveal a more authentic structure of reality, but instead, would unveil a new plane of reality that is not contradicted by virtual conditions. Therefore, we should adopt a negative epistemology that can only reveal *what the real is not*, with the dialogical plane of investigation playing a crucial role. Needless to say, technology transforms precisely this dialogical plane with the world. Technical progress alters the structure of connection that defines the intentional and peripatetic field of our existence, causing a perspectival metamorphosis that inevitably leads to a transformation of *Dasein*.

Once a technology becomes integrated into an individual's connective apparatus, the *imago mundi*, or image of the world, begins to waver, similar to the optical illusion of two figures, until the new image stabilises in the brain and the previous one gradually fades away (Changizi et al., 2008). Consequently, it is illusory to compare a largely technomediated connection with one that existed before such hybridisation occurred, as the entire image of the world, and thus of oneself, is no longer the same. The condition prior to the advent of the technology is quickly forgotten, and subsequent generations cannot even fathom what life was like before its introduction. In general, the longer the time gap between the introduction of a technology and its resulting consequences, the higher the likelihood of mistaking the induced need for the cause that prompted the use of the tool in the first place. This transformation is thus not solely about changing human performance, but it is also a profound change in experience that effectively brings out a new world. Some of the techno-cultural transformations brought about by humans, such as fire, clothing, and tool use, are so ancient that it is impossible to imagine a human condition without these technomediations.

The perspectival metamorphosis thus implies that the need to rely on technology was driven by a sense of biological incompleteness or insufficiency, caused by a changed experience. This compensatory explanation seems intuitively plausible to us because of the distant past when common tools were first used. It is important to note that even in classical times, when human beings began to contemplate their Promethean nature, they already had a substantial armamentarium of devices. It is precisely due to the decentralising effect of these devices on somatic perception that humans could observe themselves from a decentralised perspective. Today, we struggle to recognize the phenomenological experience of our ancestors without projecting our current frailties and fears onto their past. The transformation of our interface with the world has not only given us greater potential to act, but has also literally changed our lives, altering our existential coordinates. This is why it is difficult for us to understand the feelings and worldview of our ancestors.

2.5 The transformation of the social dimension

Perception also relates to self-image in social relations, introducing a dimension of life that arises from conviviality and belonging. It is within this context of social interaction that the subject's sense of adherence to the problems posed to them comes to the fore (Pierson, 2002). We often overlook the fact that ability and adequacy are commensurate with the socially shared lifestyle, and that these two parameters must necessarily evolve together. Our feelings of impropriety refer to the socio-affiliative milieu that the technologies themselves have made possible. The performative shifts produced by past technologies are hardly framed in a prospective sense, which is why we struggle to understand how people lived without the protective umbrella of certain devices. The only feeling we experience is a kind of dismay, appalled at the vulnerability borne by our ancestors. Our feelings about the latest devices, the ones we have come to know in adulthood, are of a different kind. My generation knows well that the smartphone produced a need and did not compensate for a lack.

It is now important to delve deeper into everyday life, given that each technology brings about a shift in *the individual's social interactions and engagements* (Heath et al., 2003) Each innovation leads to a transformation in social connectedness in all its dimensions, including the affiliative and participatory as well as the competitive and confrontational ones. Thus, it is crucial to consider the issue of social habits, which encompasses collective customs that dictate actions, styles, rituals, work practices, the management of private life, and leisure activities. In other words, merely analysing the potentialities gained by individuals, as if they were solitary entities facing a hostile world, is inadequate. Instead, one must take into account the various facets of our interactive existence, which are composed of social relationships, the need for belonging, and continuous confrontations to succeed or at least keep up with the cultural and historical context of social living (Balibar, 2018).

It is clear that tools are used in various fields and become an integral part of our experience. However, one question arises: are they just simple tools? I think not, considering the influence they have on the shared and relational dimensions of the body. It is reductive to see them as mere supports aimed at performance achievement, as the gradient of adequacy is defined by social perception as a whole (Ragnedda and Muschert, 2013). It is necessary to reflect on this dependence, starting with the social nexus that arises with the advent of a specific device in the sphere of participation. Objective modifications of performativity exist, but there is also an undeniable interactional effect related to two things. First, i) the attribution of a certain access to the normal relational condition of the human, where any defection is seen as a failure to achieve participatory fullness. Second, ii) the somatisation of the device, which is integrated into the body and transformed into a relational organ that makes that specific participation possible. First and foremost, there is a *sense of distance*, a feeling of *disconnection* when we are deprived of an organ-device that allows for a now fully internalised conviviality.

It is not just a matter of the habit of performance or the results it enables, but rather the plane of participation that the incorporation of the device inaugurates, which is soon felt as an integral part of a person's experience. Indeed, one of the characteristics of the human mind is to project itself onto devices that connect to the body, resulting in a sort of extension of interface ownership. Many experiments on how prosthetics are perceived demonstrate this tendency or ability to extend the somatic domain to tools. We perceive ourselves through *an interpersonal plane,* even when we engage in inner dialogue (Rosmini, 2009); somatisation is thus more than simple use. To assume that depriving oneself of an instrument that has altered the interpersonal plane corresponds solely to a performative decay is inaccurate. Instead, it becomes a failing of one's convivial dimension. When we speak of a perceived standard, we refer not only to the level of performance but also to the sense of integrity of the relational body. All systems of connection have set themselves based on that particular tool, and its deprivation alters their relational capacity.

So it is important to recognize the significant impact of shared experiences on our perception of technology as an interface extension of the body and as a means of *enhancing* participation. To dismiss it would be to disregard our social and relational dimensions, including the image of ourselves that emerges through dialogue. From a young age, we learn to navigate encounters and develop styles that become an integral part of our proxemic bubble. This is a veritable socio-ecological niche, which facilitates the relational flow. Thus we address our affective insecurities through devices like smartphones, which allow us to always stay in touch with attachment referents. Similarly, competitive motivations may manifest through luxury cars, home features or other status symbols, while affiliative needs may be expressed through uniforms, flags, or badges. The use of cultural and technological devices, thus, becomes a *transitional bond* that enables us to express our propensities with greater ease, efficiency, possibility of success, continuity, and amplification of scope (Abram, 2007). As a result, it is easy to become accustomed to connective standards facilitated by technology, and the thought of losing access to them can evoke a sense of moral loneliness.

Moreover, there is a close relationship between technomediation and the sense of control over our relationships which goes beyond just having access to others. This is an important aspect to consider: technology provides a sense of security and flexibility, giving us a feeling of certainty or, at least, greater predictability. In other words, it seems to reduce randomness, granting us greater control over our connections. As a result, removing this tool can make us feel more uncertain and vulnerable in our relationships: this certainly amplifies the perceived need for technology. However, this is just an illusion. Changing the way we communicate means being exposed to different stressors and challenges in public participation, as social media are demonstrating. Nonetheless, as social creatures, we rely heavily on interactive relationships, and anything that facilitates and amplifies these interactions can create dependence.

So when a device, such as a smartphone, becomes a means of connection, it serves a dual purpose. It enables individuals to achieve expressive goals more easily or effectively, while also personalising their social spaces. The device is not just a toy or a gadget; it is a key that allows the individual to access the feast of social interaction. The metamorphosis induced by technomediation cannot be evaluated simply by analysing the tool as a performative entity, because its action is much more global and primarily affects the subject's perception of himself in relation to others. We are also mistaken in thinking that its action merely empowers the subject: in fact, enlarging the space of relationships may not always benefit the individual, as mentioned earlier when discussing the stress of social overexposure. In addition, we must not ignore other potential negative effects, such as the assiduous use of a medium that increases control and fosters the narcissistic dimension in the relationship with others (Byung-Chul, 2018).

The perception of inadequacy we feel at the mere thought of being left without a connective technology goes beyond its practical or relational benefits, and affects a deeper level of our being. Losing access to the technology could mean failing to achieve the emotional security and social self-efficacy that it enables. However, it is overly simplistic to grant the device the role of a filter or shock absorber for social tensions, providing solutions to needs that can only find satisfaction through technomediation. In fact, the insecure person who relies on their phone to maintain constant potential contact with their emotional support system, does not compensate for their sense of insecurity, but rather amplifies it.

There are several aspects to consider regarding the dependence produced by relational media. The first, and the easiest to understand, concerns the standard or optimality that technomediation establishes. This not only leads to improvements in existing interfaces, sometimes enhancing them and sometimes extending their reach, but also creates entirely new functions, which cannot be attributed solely to an increase in performance. The human being quickly becomes accustomed to these opportunities: the technomediated function thus sets a standard of acceptability that would not be attainable with the body alone, thereby creating a perceived deficiency in its absence.

We must also view each technology as an interconnected entity within a larger technomediated system, as no technology operates in isolation. The individual does not merely use a device, but rather exists within a technomediated environment. Therefore, it is limiting to perceive each tool as a separate entity responsible for a specific transformation of the connective standard. Each invention modifies, albeit with varying degrees of transformational impact, the entire relational landscape of humanity, through processes of connection and functional emergence (Marchesini, 2017). For instance, consider the profound changes in the communicative and social spheres brought about by the convergence of telephony and computing devices. Today, we cannot view the telephone through the same lens we did in the 1970s, nor can we do so with the computer. Moreover, it is no longer possible to even separate these two apparatuses.

This is why it is not feasible to simply remove a technology: it must be substituted with an upgrade or invention that can expand the connective sphere of the original function. Technological revolutions change people's lives and do not merely enhance or extend performances already inherent in human beings (Alter, 2018). The systemic nature of the technosphere creates a niche that goes beyond a single performance to establish a lifestyle where different media contribute to an overall harmonisation. The digital revolution serves as another example which has fundamentally transformed the way we utilise various services, introducing a sense of networked integration.

Technologies not only change a person's experience but also their social participation, as they largely define participatory standards. Can a young person today do without a smartphone? No, and not just because of its specific functions, but because it's through it that relational conviviality and sometimes even social acceptance is played out. Therefore, we must not focus solely on performative aspects, because technology often becomes a veritable 'pass' to be in a group and experience a given generational dimension. Indeed, technologies characterise different age groups not only in terms of content but also of aesthetics and pace. For example, in film culture, directing and production devices and techniques change perspectives on the temporal turnover and rhythm of the narrative itself, resulting in experiences that are difficult to compare. Moreover, any technology also serves the purpose of presenting the person in their various aspects of conformity and singularity, making these apparatuses *identity devices*.

To conclude, technology has a profound impact on our connection with the world and our social participation. As a result, our self-image and related judgments of conformity, adherence, effectiveness, and efficiency are also affected. The body can never be objectively assessed, as our understanding of it is always influenced by our relational experiences. When a technology alters our perception of the body, our judgement of it is necessarily linked to the transformation of our relationship with the world around us. This leads to a sense of deficiency when we consider the possibility of losing our devices, which allow us to maintain a certain level of connectivity with the relational and convivial image we have constructed for ourselves. It is worth noting that technology not only amplifies connection and hetero-referential perception, but also makes us more exposed, needy, and open to external contamination. Furthermore, it causes a somatic decentralisation that *increases* our reliance on external elements to achieve physiological balance, which is no longer solely phylogenetic. This is not to diminish the importance of the performative aspect of technology, but to stay clear of the biases of compensation, autonomy, and disjunction.

CHAPTER 3

PHENOMENOLOGY OF THE CYBORG

Premise

In order to identify the *zeitgeist* of a particular historical era, it is useful to examine the key figures it uses to represent the human being. Imagery always relies on an *idea of the body* to express the essential characteristics of humanity. This idea is then translated into a *synthetic image* called to encompass the basic concepts of human interpretation, and in doing so, inaugurates an anthropological project (Descola, 2005). Throughout history, we have always used frames of reference to understand the human position in relation to the world: mythology in ancient times, religion in classical times, and perhaps even cave paintings in the Palaeolithic era. In the modern age, we inherited the humanistic canon, which also represents an idea-concept of the anthropological dimension (Garin, 1958). This idea-image has evolved over five hundred years through successive refinements of its initial formulations. However, the humanistic canon was challenged during the 20th century, a period that exacerbated the anthropocentric principles of humanism while also exposing its weaknesses and contradictions (Ferrando, 2020). This crisis can be seen as a transition era, induced by both the evolution of Western culture and technoscientific transformations.

An analysis of the humanistic legacy, which still impacts our anthropological understanding of the body and technology, can shed light on the transformation that has taken place in recent decades. Humanism emerged in the 15th century, drawing on classical models of *humanitas* with a continuity that was more superficial than substantial, as it was largely centred on the flourishing of Italian culture (Heinemann, 1931). The Renaissance found in Leonardo da Vinci's *Vitruvian Man* (c. 1490) its quintessential representation, which soon became associated with the humanist idea: indeed, this image is a widely recognised symbol of the human condition even today. The body inscribed in the two geometric figures – the circle symbolising heaven and the square representing earth – suggests a dual ontological dimension of human existence. Despite the earthly nature of the body, man is destined for heaven and for perfection, as indicated by the circle. The relationship between the body and the two superimposed geometric figures, where the circle dominates the square, suggests not only a cosmological union of different planes, but also transforms the human being into a reflection of the universe with a vertical aspiration (Bussagli, 1999).

The three-dimensionality of geometric forms in the Vitruvian image creates a *trompe l'oeil* effect, highlighting the body as the bearer of existential meaning, emphasising its centrality and consecrating the human body (and condition) as the universal measure of all things. The image strips away the body's technological and cultural apparatuses to expose its nakedness, with the axis of the trunk establishing centrality and the limbs creating a compass for tracing the geometry of the world. Nakedness represents an absence of natural rank, as Pico della Mirandola outlined in *De hominis dignitate* (1942), placing the body in a neutral condition that allows it to aspire to be the measure of the world. This anthropocentric and universalist intent, theorised by Pico's message and Giannozzo Manetti's *De dignitate et excellentia hominis* (1975) and outlined in the Vitruvian drawing, differs from Sophistic thought in that it no longer pertains solely to an individual perspective, but rather to the ontological condition of humanity. For Protagoras, the idea of man as the "measure of all things" implied relativism, acknowledging the subjective bias inherent in individual experience that can undermine any claim to objective truth. In contrast, humanism posits that being the measure of the world is the foundation of the universality of the human condition (Guthrie, 1969).

Humanism is rooted in the Latin classicism of authors such as Terence, Cicero, and the architect Marcus Vitruvius Pollonius. In his treatise *De architectura*, written around 15 B.C., Vitruvius saw the human body as a unit of measurement that could dictate the rules of design. Renaissance architects, such as Leon Battista Alberti, were inspired by this idea and developed *humanist anthropoplastics*, which aimed to convert natural elements to the human measure. Although *De architectura* may not be considered valuable from a literary standpoint, it was highly influential in the construction of buildings, theatres, and other urbanistic works. The fortune of this essay lies in its consistency with respect to the cultural milieu of the Renaissance. The depiction of the Vitruvian Man continued to be significant in the 16th century, with Cesare Cesariano's first edition of the treatise in Italian vernacular (1521). The Vitruvian Man, therefore, is not only a product of Leonardo da Vinci's genius but also a representation of an idea becoming a reality. The modern age was shaped by this idea, though it truly took off thanks to the European opening to the American continent, the printing press, and the scientific revolution of the 17th century.

The humanist worldview imposes a specific image of humanity that *claims to be universal*, stripping away technological and cultural apparatuses and transforming the body into a plastic entity (Armstrong, 1981). This entails that the human condition is magmatic and can be expressed in different ways, while technical apparatuses are only auxiliary. While the metric vision makes the world functional to human ends, the subsumptive image derived from the compass-man revives the principle of human autarchy and self-determination (Barry, 2002). Thus, the Vitruvian interpretation aims to *emancipate human beings* from any form of conditioning that impedes the reaching of the fated heaven inherent in their demiurgic potential. The modern age took shape by anchoring itself in classicism, electing only those examples that are functional to it, and using technoscience as a tool to free humans from the last remaining natural and theocentric constraints. The Vitruvian depiction, thus, conveys an anthropological conception that is not divorced from the vision of technology that emerged in this period (Petrosyan, 1972).

As previously mentioned, in fact, the Vitruvian Man has been widely popularised in the modern era and is still used to emphasise the uniqueness of human beings or to promote a humanistic view of existence. However, this depiction seems out of touch or inadequate from the perspective of contemporary society. This is largely due to the different understanding we now have of the relationship between the body and technology. In my opinion, what changed is the way we now interpret the body-*techne* relationship, no longer as juxtaposition or dressing, but rather in the terms of integration and infiltration (Caronia, 1985). In other words, rather than seeing technology as a separate entity, we now view it as an integral part of our being. Depicting the naked human body no longer fully captures our ontology, as technology has become deeply intertwined with our daily lives and has revolutionised the way we operate and understand the world. Recent technological advancements have even gone so far as to achieve an anthropological redefinition of our species. Therefore, we cannot fully comprehend human achievements without acknowledging the role that technology has played in shaping them.

The 20th century brought about two major revolutions that fundamentally transformed our cultural perspective and challenged the foundations of humanism, which were already under threat from Darwinian thought and continental philosophy. The first revolution can be attributed to information technology, which introduced functions and services capable of radically transforming people's lives and inviting comparisons between machine and human cognition (Reynolds, 2009). Unlike other technologies that focus on enhancing the physical capabilities, communication, and well-being of human beings, information technology infiltrates the innermost recesses of human experience, particularly in areas where humans excel (Tversky and Kahneman, 1974). Furthermore, the development of computing has led to a host of performances related to immersive interfaces, while enhancing all fields of scientific research. For instance, computing has enabled opportunities for genetic analysis, weather simulations, clinical methodologies, industrial production, and stock market transactions.

Several changes have been brought about by information technology, including: (i) the development of high-level processing, which is no longer based on simple feedback and can handle data with an unprecedented level of precision; (ii) the emergence of a comparison between human and computer processing performance, which highlights the limitations and biases of human intelligence (Art, 2022), as well as the potential of AI; (iii) the establishment of the Web, capable of implementing the globalisation of data, accelerating and facilitating the circulation of news, creating connection systems such as social media, and triggering accessibility to personal information and profiling of individuals; (iv) the emergence of the digital dimension, which has unified a large part of cultural experiences, promoting a different relationship between the subject and space-time, inaugurating, in the sign of immersion, a second, somewhat alternative world; and (v) the development of artificial intelligences endowed with self-programming, that is, machines that are no longer expressions of human programming but the outcome of an evolutionary process (Gawdat, 2021).

The other major upheaval can be attributed to the development of biotechnology, which has created a kind of applicative horizontality of life. This is evident in the fields of genetic and protein engineering, which have freed species-specific traits from their natural location and made the gene stateless, allowing it to migrate and configure itself within a new genomic matrix. The breakthrough came with the discovery of the structure of DNA by James Watson and Francis Crick in 1953, which allowed us to delve into the heart of life by deciphering protein synthesis and the mechanisms of phenotypic translation of traits. Relating the information encoded in the genome to the morpho-functional expressions of the body has shown us the possibility of redesigning the human being to enable them to perform tasks not present in their phylogenetic heritage. The discovery of restriction enzymes, which act like molecular scissors, allows individual genes to be cut out, paving the way for genome editing. For example, the Crispr technique developed by Jennifer Doudna and Emmanuelle Charpentier has made gene correction and trait determination feasible (Esvelt and Wang, 2013). Gene therapy is thus becoming a reality, opening up a range of applications that extend beyond the field of cure.

Procreative biotechnology, likewise, has revolutionised the canons of human reproduction by allowing intervention in the earliest steps of the organism's constitution. In vitro fertilisation, following superovulation, allows for the evaluation of individual embryos and the selection of which ones to implant, with clear eugenetic overtones. Furthermore, embryo manipulation techniques, such as splitting and fusion, have dissolved species barriers, giving birth to chimaera animals previously confined to mythology (Illmensee and Levanduski, 2010). In the 1980s these technologies, initially tested on other species, initiated the possibility of intervention on the human embryo, inevitably raising major ethical questions. Then, Ian Wilmut's cloning of Dolly the sheep in 1996 broke down the distinction between somatic and germ cells through the process of nuclear transfer, making cloning possible and bringing to fruition the human dream of self-replication (Barresi and Gilbert, 2020).

The convergence of computer science and biotechnology has led to the development of biosynthesis, which removes the barriers that previously limited imagination in the field of science fiction. The cell can now synthesise molecules with extreme precision and specific functions; all that is needed is to program it to do that particular job to make it a factory at our disposal. Biosynthesis has enabled many conversions, some of which involve the war industry. For example, hybridisation in the microorganic sphere can turn viruses into weapons of mass destruction. Unfortunately, this seems to be the new scenario between the great powers of the planet, after that of nuclear bombs (de Clercq and Kern, 2003). Although biotechnology may seem to have less of an impact on people's daily lives, pharmacogenetics, particularly the increasing use of psychotropic drugs that can enhance the action of certain neuromodulators, is an area not to be underestimated. The possibility of directly intervening in the mechanisms of living organisms and modifying their performance has caused a real shock in the imagination, not to mention the potential of bioinformatics (Attwood et al., 2011).

Oppositional dichotomies, such as nature versus culture, invention versus discovery, and human versus non-human, which were once the foundation of Western thought, are progressively losing their meaning. Technology is increasingly resembling a virus that can infect the body cell and reprogram its functions (Marchesini, 2021). As a result, a new concept of tool is emerging, in which the instrument, despite its name, is gradually losing its instrumental qualities. The medium is no longer entirely at the body's disposal; on the contrary, it seems to invert the traditional canon: the body, by accepting it, must adapt to the device. Technology now appears as a parasite that penetrates the body's pores and embeds itself in the body's connections, modulating its functions. Hence a *new infiltrative concept of the instrument*. This is the case with endodermal chips, nanotechnology, gene therapies, and neuro-reuptake inhibitors. It is evident that these devices are markedly infiltrative, and one cannot reduce this qualitative leap to a mere heightened awareness of the impact of technology on the body. Rather, we are faced with an emergent process resulting from the convergence of extremely productive research areas with significant implications for the body's dynamics (Renner, 2013).

In conclusion, while embodiment has always been present in devices through a multitude of somatic reorganisation processes, the significance of these techniques is now more apparent due to their blatant infiltrative nature. This has given rise to a new paradigmatic figure, *the cyborg*, which is a somato-technological hybrid and a new model for the human condition (Haraway, 2016). The phenomenology of the cyborg disrupts the principles of intentionality by not just introducing new contents of predicative reference, but also altering the subject's perspective, nullifying any demand of balance and immutability. It is from this new awareness that we must begin to evaluate the predicates of the human, which are changing not only because these technologies have profoundly pushed the barriers of the body, but more importantly because they have allowed us to view the predicative process itself in a new light: no longer as an emanation of intrinsic contents of human essence but rather as an emergence of the relational capacity of the human.

3.1 The relationship between *techne* and body

In our era, the cyborg embodies the plastic representation and, I would say, the very poetics of predicative hybridisation. As classical metaphysical barriers are being overcome, the cyborg encourages us to view entities not in isolation, extrapolated from their milieu, but from an *eco-ontological* perspective that emphasises their *interconnecting structures* (Bateson, 1976). These structures are the true predicative drivers of the subject, not some illusory essences that can be extracted in their purity. As a result, the somatic dimension becomes the embodiment of this dialogic-creative tension, capable of generating novel forms and related functions through contamination (Hochberg et al., 2012).

The body undergoes a state of magmatic ferment, marked by mutation and transition, where it is increasingly fluid and reorganises itself on new morpho-poietic lattices. Information technology and biotechnology highlight the *somato-technological connection*, illustrating that the perceived separation between instrument and organism not only involves psychological and projective perceptions of function, but also deeply impacts tissue organisation and metabolism (Velliste et al., 2008). The impact of new technologies on the physiology of the body is evident in their immersive and infiltrative nature, stemming from their totalising and pervasive actions. While the analogue use of various tools previously led to distinct experiential processes, resulting in plural metacognitive exercises, the digital media's unification of these experiences always trains or reduces the same functions, thus intensifying its effect.

Although devices have always had consequences in terms of somatic redefinition, their impact today is more pronounced and evident than ever. While tools have always interfered with organic functions, the influence of digital devices is more constant, potent, and targeted. The rise of digital technology has created an alternative world where individuals can shift seamlessly from information-seeking to entertainment, from social interaction to constructing their own online persona while remaining connected to the same media. People now spend an estimated eight hours a day on the internet, which results in a metamorphosis in their daily lives that has physiological repercussions on all bodily systems (Greenfield, 2015). This immersive permanence can be compared to the living conditions of factory workers in the 19th century, with the critical distinction that machines no longer impose physical labour and risks of injury. Nevertheless, these habits can still have debilitating effects on the body. While standing still for many hours produced biomechanical alterations in the body in the past, the current habit of sitting with the neck stretched forward has led to spinal restructuring with significant negative impacts on health (Cregan-Reid, 2018).

The body is always moulded by the technology or by the practices the latter entails, so we can say that the tool becomes embodied in the somatic dimension, even when it appears external to the subject. As the body stretches along the technical medium's coordinates, it blends with the tool and experiences new existential dimensions. The cyborg is a point of confluence and reciprocation: to be inhabited by technology is to inhabit its spaces. Digital technologies are a clear example of this, profoundly transforming the very meaning of experience. Every time a person accesses the Web, they are being watched, and every online action leaves a trace, altering spatio-temporal values. The cyborg expresses an out-of-body presence that can turn into a post-body existence, into an active life within the Web even after death (Sisto, 2018). I speak of an ecstatic dimension of the digital to emphasise the ontological metamorphosis that this technology has produced, transforming bits into a new first principle and positionality into a different approach to the interpretation of reality. These devices have high connective potential and strong evolvability content, precisely because of the hybridising aspect they imply. The cyborg participates in conditions that are otherwise precluded to the individual.

Machines are becoming increasingly interconnected, leading to emergent processes that are difficult to predict, yet these phenomena are already unfolding before our eyes. Digital devices can create direct interfaces between the body and machines, and so we are excited by the prospect of being able to move a device with our thoughts. However, we fail to consider the possibility that the opposite may soon occur, with our dispositions being controlled through Internet inputs (Gawdat, 2021). The use of digital social networks is changing the very nature of social participation, and neuropsychological studies have already highlighted some relevant consequences, after just two decades of Facebook's existence (since February 2004). The underlying principle is that of permanent connection, which is made explicit today through a headset, but may soon be fully integrated within the body. Yet, even within our present reality, mediated through smartphones, we cannot ignore the fact that the concept of permanent connection already implies a hybrid dimension of existence (Floridi, 2014). We are already existing in a state of hybridisation, as there are no longer any filters between us and the world, and we cannot turn off our cell phones if we wish to participate in this new form of *ecumene*.

Gaining new awareness of the effects of technology on the body through the image of the cyborg may, for some scholars, not only represent an interpretive code but also a blueprint for the future of humanity, just as the Vitruvian man did in the 15th century. The transhumanist movement, for example, envisions a posthumanity that is completely liberated from the constraints and vulnerabilities of the body through the mutagenic and transcendent power inherent in technology (Bostrom, 2018). Just as the Vitruvian man undresses to reveal his essential metric-subjunctive dimension, the transhumanist cyborg hides the body to the point of annihilation to express its total eradication from the phenomenology of life. The result is an enhanced image, a kind of *anthropocentrism without humanity*, which, in some ways, echoes the ascensionality and disconnection from nature that are emphasised in humanism. This vision is indicative of a feeling of the contemporary age that fascinates some and horrifies others, even in its imaginative character. I believe that the figure of the cyborg is useful in interpreting the human condition, which inevitably has a hybrid dimension that is made even more complex today by new technologies. However, the cyborg does not represent the power of technological humanity, but rather its fragility, its increased exposure and dependence on the world, and its heightened sensitivity (Haraway, 2016).

Let me make myself clear: I believe that human fragility is a very powerful quality, and to recognize this condition is to acknowledge a shared destiny with otherness, avoiding the trap of narcissistic solipsism. Acknowledging our hybrid condition does not mean achieving humanist ascension, but rather understanding the insubstantiality of the anthropocentric project. Recognising that the body is a malleable entity shaped by its developmental niche means abandoning any claim to autonomy and existential self-determination. The cyborg serves as a warning to maintain critical awareness about human beings and the variations that technology can produce. Every innovation presents opportunities, but also problematic issues that require humility and the use of all available conceptual tools. As we increase our operational reach through technology, we must also develop ethical reflection and avoid both technophobia and techno-enthusiasm (Jonas, 1984).

The question of the phenomenology of the cyborg requires us to move beyond the conventional understanding of tools and ergonomic readings and instead focus on *technophysiology*. We are currently witnessing a significant anthropological revolution, but we are attempting to comprehend it with the tools of traditional humanism. This is a mistake, as it assumes a stable human condition, with inherent qualities, goals, and dignity, that can simply *utilise* technology. However, this ontological entity is gradually disappearing – provided it ever existed in the first place. The cyborg urges us to embrace a vision that is less focused on celebrating humanity and more inclusive of the biosphere. Bringing human manifestations back to a relational dimension means accepting predicative co-factorship, which calls for a cautious and moderate approach (Schaab, 2013). Such a transformation requires us to rethink traditional models of understanding the human condition, which are still rooted in essentialism (Fuchs, 2005). The cyborg, instead, highlights the value of contamination and its productive potential, as well as the sterility of principles based on purification and disjunction. In terms of ontopoietic ritual, the cyborg inaugurates an inverse canon that is based on relational variation instead of the alchemical extraction of an alleged essence. Therefore, the cyborg reminds us of the value of *convivium*.

The transformation of the relationship between the body and technology, in terms of *technophysiology*, is thus well exemplified by the image of the cyborg, which, unlike the Vetruvian Man, cannot be traced back to any prototypical model. Moreover, unlike the Vitruvian man, the cyborg cannot strip off its devices because they are part of its somatic dimension. It is important to note that this change in perspective is not imposed from above but is a bottom-up transformation that is ingrained in the daily practices of contemporary societies. The cyborg is not a product of intellectual fantasies but rather the form that humanity has taken in expressing its everyday life (Chislenko, 1995). The challenge is to become aware of it and extend that awareness in retrospect, because we have always been hybrid entities. Rather, today the prominence of this hybridity can no longer be ignored, because we now live in a constant state of connection and phleboclysis with technology. As previously mentioned, digital tools have become omnipresent intermediaries that accompany us in every moment of our lives: it is natural that they have a much greater impact on the physiology of the body. A similar role can be attributed to psychotropic drugs – inhibitors or stimulants of certain neuromodulation functions – whose consumption has increased to such an extent that their role has shifted from curative support to a form of doping (Breggin and Breggin, 2014). As a result, our bodies are constantly under the influence of these modulators, which affect the functional regulation of organs, differential development of tissues, decoupling of functions, co-optation of an apparatus to new functions, and dependence of the organism on external contributions.

Despite the traditional aim of ergonomic design to make devices more body-friendly, it is often the body that must adapt to the tool. This is not due to a presumed dominance of the tool over the body – although technologies often do become cogent for reasons of social acceptance – but rather because the body is far more adaptable than any device produced by humans so far. However, this does not always have positive outcomes, and the pervasive nature of technology is also revealing problematic effects on the health of the body. Autoimmune diseases are on the rise, and there are more frequent alterations to the menstrual cycle, as well as the spread of pathologies such as endometriosis, constant reduction in sperm count in Western countries (Swan, 2022), postural impairments, and attention disorders, just to name a few examples. These effects are significant and cannot be ignored. The body infiltrated by technology has been experiencing a phase of crisis in these decades because of its apparent fragility compared to the hardness of metal and silicon or the immediate effect of a psychotropic. This body feels precarious compared to the solidity of the things around it, and at the same time it is often accused of obsolescence and vulnerability.

The pervasive influence of technology in contemporary society has brought about a deep-seated anxiety regarding our organic condition. This is the first time such an anxiety has been so strongly felt. Despite being incorrect, there is a perception that our bodies are obsolete compared to the potential of technology (Macrì, 1996). A few examples can illustrate this point. The body has a continuous and systemic structure, unlike machines which are composed of modular entities that can be easily replaced if broken. The body is also made of delicate material which can be easily damaged, making it much less robust than the chassis of a robot. Furthermore, the body has physiological demands that can limit its versatility, such as the need for homeothermy, large organic functions, hormonal cyclicities, and ageing. In contrast to information, which can theoretically travel at the speed of light, the body is constrained by the space-time dimension. Information technologies have the ability to interconnect all machines and create a collective intelligence similar to that of a hive, which is impossible for the individualistic human brain to achieve (Wheeler, 2012). Most of all, the body has an inevitable expiration date that cannot be avoided, only delayed.

The body is a product of evolution, meaning that it has undergone long periods of adaptation through exaptation, so each organ and function is a product of the resources available to that lineage and the adaptive demands placed upon it. This condition of compromise has resulted in vestigial appendages in the body, as well as a functional organisation that may not always perform optimally when evaluated based on specific performance principles (Gould and Lewontin, 1979). Let me clarify: I am not suggesting that an organism is in any way inferior to our technological apparatuses, which are actually very far from the complexity of a single cell. Rather, I am saying that we are the victims of this mistaken perception, for a multitude of reasons, including anthropocentric fascination and the pride that machines inspire as our creations. I would not underestimate the pygmalion-like dimension of anthropoplasty, given the arrogance of human beings! That being said, there is no denying that devices have their strengths: objects, buildings, and machines can outlive our existence, surpassing us in time, and this greatly influences our perception of them.

Our evaluations, however, fail to account for numerous other qualities that render the organic far more malleable, adaptable, integrated, and innovative in generating new functionalities than today's technologies. In contrast, machines are characterised by their relative controllability, precisely because this is the task we have assigned them (Hughes, 2004). Were a machine to lack standardised performance requirements, controllability of output, reduction in chaotic effects and noise, and reliability in operation, it would cease to be a machine. Still, human beings do feel vulnerable in the face of metal's hardness, lagging behind the speed of motors, left powerlessness before mechanical strength, and deficient compared to a computer.

Indeed, computer science has radically altered our perception of machines, as evident from the very concept of artificial intelligence, in terms of both its performance and self-programming capabilities (Mitchell, 2019). AI has played a significant role in challenging the anthropocentric view that placed humans at the top of the cognitive hierarchy – the Deep Blue software beating world chess champion Garri Kasparov is a small example of this. The advent of processing machines capable of problem-solving and learning, and their ability to evolve beyond the programmer's instructions, has enshrined the obsolescence of the human designer. This decentralisation of humans in their most boasted faculties represents a significant rupture, as illustrated by the project to enhance or amplify human intelligence through the incorporation of silicon devices. The cyborg is therefore an upgrade, a kind of *human 2.0* capable of rectifying the last remaining flaws in the system. Information technologies have long ago exposed the intellectual limitations of human beings, previously believed to be the pinnacle of intelligence, by bringing out the numerous *cognitive biases* that characterise them (Buss, 2005).

The impact of technology on the human body has been highlighted not only by the significant advancements in the two previously mentioned revolutions and their impact on daily life and imagination, but also by the significant biological discoveries of recent decades. Our understanding of living processes has given us new tools to control the physiology of the body, allowing us to modify somatic performance (Francis, 2015). In other words, the cyborg is not just a product of technological infiltration, but also the result of a better understanding of the somatic dimension and how to carry out grafts successfully (LeMoyne, 2016). The cyborg, therefore, results from a convergence of different fields of research that has found a renewed interest in anthropology, with the body as the ground for comparison and operation. Our better understanding now extends to embryonic development, gene translation, metabolic influences on mental health, response mechanisms to environmental factors, pathways for bionic grafts, and the activation characteristics of neurons.

Indeed, we should not forget that the 20th century witnessed a remarkable increase in our understanding of biological mechanisms responsible for the production of phenotypes, including studies on the evolution-development relationship and epigenetics. Biologists have revealed the potentiality of genotypes, which inherently exhibit polyphenism and result in the phenotype of an individual being dependent on external variables. In nature, any alteration of a niche produces a metamorphosis of the phenotype, altering the phylogenetic trajectory of the species (Gilbert and Epel, 2015). Today, we recognize that technological apparatuses are, in fact, a niche, and any changes in them alter the human phenotypic plane. Physiological research has also yielded significant contributions, particularly regarding the interaction between the neurobiological system and the two auxiliary knowledge apparatuses, namely the endocrine and the immune systems. Novel techniques of electrostimulation and neural imaging have provided detailed information on the anatomo-functional basis of psychological conditions and how hormones and cytokines influence the activation of various neuromodulation pathways. In the past few decades, pharmacological studies have generated substances that can manipulate the most important physiological processes, illuminating the body's permeability to external interventions, even in traditionally obscure aspects such as the biochemistry of affective states (Carlson, 2013).

What is most striking in this metamorphosis of the technosphere is the convergence of the two revolutions, leading to a change in the status of living organisms. An example of this ontological assimilation is the construction of increasingly infiltrative and wireless interfaces between organisms and various informational media, allowing the praxic dimension of typing to transition into the more direct and transcendent dimension of thinking (Romo et al. 2000). Indeed, while any microchip graft within the body can facilitate direct interaction with machines through simple connection, interfaces that are more aligned with the sense-motor apparatus can enable thought-activated performance (Talwar et al. 2002). The notion of machines infiltrating the sacred somatic recesses of the body may seem appalling and scandalous, but we should remember that this technology can address specific sense-motor deficits resulting from accidents or degenerative diseases.

The same is true of biotechnological perspectives, often stigmatised as an expression of human hubris, claiming to modify a nature that is considered perfect. However, this view forgets that, if we accept evolutionary thinking, perfection is simply not given in nature. Conversely, some biotechnological processes that may seem controversial, such as gene transfer from one species to another, are commonplace in the natural world, especially among microorganisms. Moreover, the development of epigenetics has already blurred the somatic-germinal barrier, making it increasingly difficult to define limits, except for a vague precautionary principle that is becoming increasingly precarious in its application. The problem lies in our perception of the body, which is still considered sacred and inviolable, and any action not aimed at restoring compromised function is viewed as dangerous or illicit. However, this boundary is becoming increasingly blurry, especially when considering the advances made by cosmetic surgery. Indeed, although the distinction between somatic and germinal intervention still holds, it is not always obvious that somatic modifications are unacceptable and germinal interventions are illicit.

Certainly, the notion that there is an insurmountable divide between the body and technology is nothing more than an old cultural legacy that stems from a misreading of somatic organisation. In reality, each technological innovation infiltrates the ontogenetic processes of development, shaping the development of the neurobiological, lymphatic, and endocrine systems in response to adaptive counterparts. *We have always been cyborgs* (Sorgner, 2022): even the use of chipped stones led to differential development through exercise, increasing certain somatic areas at the expense of others and altering selective pressures to favour mutations that reduced the splanchnocranium to benefit the neurocranium. Thus, every technology produces somatic transformations that shape our flesh. Still, it is understandable that the direct grafting of chips onto cybernetic interfaces or connecting the neuro-biological system with silicon implants would have a different emotional impact. It is important to recognize, however, that machines are also evolving at a rapid pace. This is thanks to the development of increasingly sophisticated operating systems and movement mechanisms, as well as the study of biomechanics, which can provide calibrated models and solutions based on millions of years of adaptive selection (Kelly, 1994).

It is beyond doubt that if we are to truly understand the ongoing changes in human-technology interactions, we must first change the cultural paradigm through which we interpret the role of technology in human ontology. The humanistic tradition sees technomediation as complementary, leading to the inference that there is a juxtaposition rather than a hybridisation between the two terms. In this paradigm, the concept of *techne* takes on certain characteristics that make it difficult to perceive modifications of the body except in the most conspicuous forms of grafting. According to this view, technology: i) functionally supplements the human by providing missing endowments; ii) enhances certain inherent attributes by expanding their field of operation; iii) conforms to the body, which informs technology and does not undergo it; and iv) serves as an instrument to achieve human ends. Ultimately, the Promethean myth of the incomplete human being compensated and completed through technology still influences the way we think about this relationship. This is a serious error, both from a biological standpoint and in our understanding of the technopoietic interface. The future will undoubtedly bring about increasing hybridisation between humans and machines, with entirely unpredictable outcomes. Yet, this history of contamination has been underway for millennia.

3.2 The cyborg: a new model for the human condition

In 1985, Donna Haraway's essay "A Cyborg Manifesto" was published in *Socialist Review*. In it, she presented the figure of the somato-technological hybrid as a symbol for the overcoming of all forms of dualism. In reality, this view only blurs the traditional categories of human, animal, and machine, which were the metaphysical gaps underlying the classical paradigm, already challenged by the waning of humanism. Undoubtedly, Haraway's text represents a significant turning point in the interpretation of the relationship between the organic and the technological, as well as a radical critique of ontological anthropocentrism (Haraway, 2016). However, upon closer inspection, the text merely reflects a feeling that was gradually taking hold in the cultural climate of the late 20th century.

Since the emergence of cybernetics, the Cartesian approach, which used the machine as a model to explain the organism, underwent an abrupt reversal (Mc Culloch and Pitts, 1943). Now, the design of machines increasingly sought inspiration from the biological model, or at least did so explicitly, by plundering biomechanics and feedback mechanisms to create devices capable of performing specific tasks and modulating functions intelligently (Wiener, 1948). *Biomimicry*, in its various forms, undoubtedly experienced a new surge during this period, as it became clear that natural selection had gradually – though over very long periods of time – devised particularly ingenious solutions that could be studied and imitated (Fournier, 2018).

The cyborg thus signifies a fundamental shift in the way we understand the human condition, diverging significantly from the previous humanist paradigm. This new model represents a cyborg phenomenology that goes beyond a mere description of a body invaded by technology, and instead alludes to deeper meanings and implications related to ontology (Garcia, 2010). The Vitruvian Man, who undresses himself and posits his own body as the measure of the world, links his somatic condition to his ontological condition, where the latter depends on the former. This emanative view of predicates implies that what is manifested by the human being is directly linked to the qualities of the body itself. The human appearance is seen as synchronous with a certain bodily condition, and everything else is considered ontologically irrelevant – hence the Vitruvian Man's undressing. Conversely, the cyborg *emphasises* its dependence on external factors, the hybrid nature of its predicates, and therefore its inability to undress and regain its primal ontological condition, because its appearance is always a correlation (Longo, 2003).

The cyborg is thus the result of a changed mindset that aims to bridge the gap between *bios* and *techne*. In a way, it celebrates a renewed fusion of nature and technology, which find new forms of synthesis. The identity of the cyborg is located in a logic of magmatism and contamination, breaking down previously parallel domains. This leads to a metamorphosis and a technomediated reconfiguration of the human, which goes beyond simple ontogenesis and is rather ontopoiesis. The human condition seems to leave behind the safe harbour of morphological stability and embark on a journey of continuous change, hinting at a becoming that extends beyond the human. The cyborg is driven by existential restlessness and seeks to embrace the technological revolutions of the 20th century, almost dissatisfied with the slow pace of phylogenetic processes that constrain the species within an antiquated phenotypic shell (Anders, 1980). The cyborg desires constant technological upgrades, as if the sensory transduction system were too narrow to satisfy its hunger for new horizons. Ultimately, the cyborg dreams of a broadband that enables more powerful and versatile access, capable of handling and conveying more information and speeding up its transmission in a multitasking dimension (More, 1995).

To comprehend the phenomenology of the cyborg, it is crucial to adopt a constructive or ontopoietic perspective of the human condition. While the Vitruvian Man utilises the body to demarcate the geometries of the world, the cyborg employs the world to construct its own existential geometries. The ontopoietic principle that governs the cyborg's manifestation and its intentional structure lies in the construction of a niche, which is not external to the body but an integral part of it. Therefore, it is impossible to understand the meanings implied or alluded to by this paradigmatic figure without exploring the principles of niche construction that relate to human practical tendencies. However, once again we must depart from the humanist perspective that purports to explain the human capacity for adaptability solely based on the naked body. According to cyborg logic, we cannot derive human manipulative skills directly from the conformation of the hand, even though it is certainly essential to consider its morphology. Rather, we must consider the dressing of the hand, from the interactive glove to the subcutaneous chip. The tool that characterises us is not the one we employ to act on the world, but the one we utilise to construct another device (Leonhard, 2016).

Two factors are at play in this kind of meta-technopoiesis: the somatic factor, which is defined by the hand's sensorimotor and articular qualities, and the instrumental factor, based on which the hand seeks to fully adhere to the constructive task. If we examine the technique used by early hominids to create a chopper, we will find the foundation of this principle: using a stone as the primary tool to strike another stone and chip it, so as to obtain the final and secondary tool. This is a basic craft model, a tool for making tools, which finds its clearest explanation in the lathe (Clifford, 2018). Meta-technopoiesis can no longer be directly traced back to the hand's conformation, even though the importance of the hand in adhering to the primary tool remains clear. However, the most interesting aspect is that the entire practognostic organisation, not just the hand, must reconfigure itself when a tool is created through another tool. It is not just a matter of acquiring kinaesthesia but of reconsidering the very dictate of manipulation, i.e., the way of conceiving niche intervention. One of the aspects that come most into play in this process is the cognitive factor of coordination: in fact, it is necessary to refine the precision of eyesight, which must now work not only with the hand but also with the hand-instrument hybrid.

It is clear that technopoiesis has always relied on eye-hand coordination, from the earliest stone tools to the current act of typing on a touchscreen monitor (Borin, 2005). The ability to visualise the part to be worked on is linked to the movement of the hand, which represents its potential perspectives, allowing the meta-tool to be rotated, tilted, inverted, or scrolled. Depending on how the meta-instrument is manipulated, the potential contents of the final product change. When we watch a science fiction movie where the protagonist interacts with a remote monitor, we can glimpse the same coordination matrix that allowed our phyletic line to develop amygdalae and bifaces. Practognosis has always been at the core of our niche construction style, but in *techne* it refers to a somato-technological hybrid, not a bare hand. The cyborg emphasises the impossibility of undressing like the Vitruvian Man, branding it as ontological anthropocentrism. The Vitruvian Man, in fact, claims that humanity is derived from the naked body, considering technology as a mere aid, useful but not indispensable for ontological analysis. On the contrary, to understand humankind, it is necessary to acknowledge the importance of somato-technological hybridisation (Terrosi, 1997).

In this view, the human body is not seen as incomplete or imperfect but rather as possessing such redundancy that it can develop an extensive range of virtual possibilities. Human beings see everything around them as not only something to be enjoyed but also something to be incorporated. At the same time, the world can be somatised by us because of this redundancy, which enables us to take on organisational information from outside. The cyborg exemplifies this inherent propensity to view everything around it as a potential somatic extension, a continuity of its own flesh. There is a very close relationship between seeing and touching in us, so much so that when an object is far away, our hand reaches out: this is the earliest form of indication developed in childhood, also encompassing the yearning for incorporation that is innate in human beings. Seeing means assigning an affordance to an object (Gibson, 1979), but also imagining hidden possibilities, so every object is metaphorically treated as a fruit to be peeled. To understand technomediation, it is necessary to refer to practognosis and the meaning it has taken on in human history.

In my discussion of the specialised morphology that characterises our species, I have focused on the anatomical changes in the foot, spine, pelvis, and femur that allowed for bipedalism. Similarly, one can highlight the importance of the manipulative ability facilitated by the hand, a remarkably complex organ with numerous sensory receptors (Hashimoto et al., 2013). Subcutaneously, the hand contains tens of thousands of receptors with a wide range of functions and variations in their reception: i) Merkel cells enable us to recognize shapes; ii) Ruffini corpuscles allow us to sense movements on the skin; iii) Pacini corpuscles enable us to feel pressure changes and vibrations; iv) Meissner corpuscles inform us about the texture of surfaces; and v) free endings provide us with information about temperature, itching, and pain sensations. The hand, with its special prehensile conformation and performative ductility of the different fingers, has enabled tool-making and technical activity to a greater extent than any other organ. These characteristics, in fact, are already present in primates, but not to such a refined degree (Wilson, 1999). However, we should not consider the hand as a simple organ that can perform many functions, as it actually makes use of the three joints of the wrist, elbow, and shoulder, as well as a highly refined precision retinal system and a great capacity for eye movement. This ensures stability through vestibular signals and keeps the gaze fixed on a point through vergence movements, while also monitoring the entire visual field to locate new objects of interest to the fovea with saccadic movements (Larsson, 2013).

When examining the practognostic capabilities of human beings, we find that operational execution develops from the involvement of innate features, such as the particular brain conformation of the visual areas. However, these features evolved precisely because of a technomediated niche that can be traced back as early as the australopithecines, who already exhibited prerequisite technical properties, as observed in chimpanzees (Heesy and Hall, 2010). While the hand was originally prehensile due to the arboreal life of primates, its conformation manifested an evolvability that allowed it to become an armed hand that first acted on the world, and later through it in a meta-technical way. The cyborg, as a useful paradigmatic model for understanding humanity, indicates that the latter aspect is what most connotes the human, rather than the use of tools. However, this also makes it clear that understanding the human solely through recognition of the body is impossible, because the human dimension lies in hybridisation. In the present day, the cyborg not only signals this but also foreshadows changes for the near future that are capable of increasing somato-technical fusion. The metamorphosis of the contemporary cyborg moves precisely in this direction, transforming from an epistemic model to a sometimes excessive project: even the hand appears to suffer from obsolescence in this context.

The cyborg, in fact, presents the possibility of more versatile joints capable of accommodating various types of prostheses. For example, a polyfunctional meta-carpal could act as scissors, forceps, microtomes, or other specialised tools, in a way that is reminiscent of Tim Burton's film *Edward Scissorhands* (1990). The cyborg longs for prostheses that can perform currently inaccessible tasks, such as microsurgery, direct computer input, or miniaturisation of devices. In essence, it wishes for an attachment that can be adapted to multifunctional grafts. The manipulative conception of the cyborg challenges the idea of the hand as the ultimate practognostic structure. The desired *somatic fusion of the tool* is increasingly viewed as something that can bypass the hand altogether (Gray, 1995). And this is only the first step towards non-manual manipulation, where the brain can interact directly with various devices. The cyborg's projection of manipulation bypasses all forms of somatic-practognostic mediation, in view of acting directly on the world through thought thanks to a transductive chip implanted in the brain capable of wirelessly controlling various machines.

The cyborg therefore embodies dynamism, restlessness, and a *compulsion towards innovation*, as it is its own self-project and upholds the principle of updating morpho-functionality. To truly comprehend the phenomenology of the cyborg, it is not sufficient to only examine the possibilities provided by technoscience; we must also contemplate the conceptual transformation of the organic within the current cultural milieu (Thweatt-Bates, 2016). We have transitioned from a deeply essentialist and anatomo-functional stability-centred conception of the body, which was still prevalent during the first half of the 20th century, to a view that is based on the mutation and transformation of morphotypes, the continuous process of readjustment and ontogenetic co-factorism, and the ecological dynamics of coevolution and symbiogenesis (Margulis, 1998). Life is now depicted more as a process than a substance, analysed in its fluctuating and imbalanced meanings, so much so that even the organism is likened to a stream, river, or fire – constantly changing. Ultimately, the taboo of the body's intangibility has fallen, along with the notions of perfection, the sacredness of form, and the ideology of a stable nature acting as an index of order.

The cyborg embodies this state of constant change, *ontological meta-stability*, and the need for continual self-transcendence. As a result, essentialism also comes under scrutiny in this perspective dominated by mutagenic and hybridising events: form arises from chaotic processes and from the bottom up. It is in this cultural climate, characterised by *fluid identities* (Miglietti, 2008) and infectious processes, that the cyborg finds its place as an epochal model. In this sense, the comparison with Leonardo da Vinci's depiction of the Vitruvian Man helps us understand the heuristic significance of the cyborg and the fascination it exerts. The cyborg represents a fusion between the body and technology that brings about significant shifts in the concept of humanity, including: i) the idea of an updatable operating system; ii) the supposed programmability of the organism's functions; and iii) the notion of a cofactor identity.

The 1980s marked a turning point in innovation, not only influencing the spirit of the times but also scientific research. With findings in genetics having been consolidated, this period saw a frenzied search for new applications in medicine and agro-zootechnology. Biotechnology brought with it a period of uncritical optimism, as the first transgenic experiments (Brinster and Palmiter, 1982) began to capture people's imagination. There was talk of genetically modified organisms for studying diseases, developing new drugs, increasing agricultural and farming productivity, and creating transgenic organ banks. The study of embryo and germ cell manipulation, known as procreationics, also yielded results that would shock the imagination. Despite limited practical applications, they influenced the cultural climate that economist Jeremy Rifkin dubbed "algeny" in consonance with mediaeval alchemy (Rifkin, 1983). In this new horizon of *bios*, boundaries fell away one after the other: it was found that a genetic trait from one organism can be transferred to another, the nucleus of a somatic cell can trigger an embryonic process, and embryos from different species can be fused together. However, what was most shocking was the realisation, which was not slow to emerge, of us humans being not only agents but also recipients of change.

The cyborg owes a significant debt to psychedelia and the dream of expanding consciousness. Timothy Leary promoted forms of ecstasy and immersion through hallucinogens in the 1960s, and his approach had a decisive influence on cyberpunk, the conception of virtual reality, and the design of interactive monitors. In this cultural climate of transcendence and dreamlike hyperbole, drugs became the means of leaving behind the ordinary dimension, reprogramming one's psyche, and gaining access to new horizons. It was this context that generated the idea of transforming the computer from a passive, cumbersome, and distant tool – the classic data processor – into an intermediary for escaping from a reality felt as alienating and oppressive. At the same time, with the advent of personal computers, such a possibility began to be within the reach of the general public: psychedelia was thus projected onto an instrument or gadget (Rheingold, 1992). That was an electrifying shift, because these technologies seem to enter deeply into people's experience, colonising certain recesses of their minds such as psychological mechanisms and information processing, which were previously thought to be strictly human. Whereas appliances had changed the way of living but still remained external to the body, from then on, things seemed to change irreversibly.

The cyborg, which Haraway presents as a model, shows us how the convergence of information technology and biotechnology was on the verge of opening up alternative perspectives on the role, or fate, of technology itself, as early as the 1980s. Suddenly, the infiltrative presence of technology took on substance and evidence, and *techne* was close to forever disrupting the warp of the human body that celebrated its autonomy as *copula mundi* in the Vitruvian Man (Kristeller, 1943). The cyborg, instead, is a pierced St. Sebastian that, nevertheless, has opened its porosities and surrendered itself to penetration not for the sake of martyrdom, but to continue floating in the stormy sea of a dystopian future that has sanctioned the obsolescence of bare flesh. The more technology infiltrates the body, the more the flesh abandons its compactness and follows coordinates imposed by the device.

The imagery of the last decades of the 20th century is charged with new protagonists who have made their hybrid condition a disenchanted normality. They can no longer be classified as deviants, monsters, doppelgangers, empty shadows, or comprimarios because, at this point, it is clear that they embody a new polyphonic protagonism that has abandoned the claim to a canon.

Films such as Ridley Scott's *Blade Runner* (1982) introduced a new vision of the replicant, no longer seen as a mere copy – as could be found in Fritz Lang's *Metropolis* (1927) or Don Siegel's *Invasion of the Body Snatchers* (1956) – but as a being with its own experiences and perspectives, capable of integrating the fabric of humanity. The replicant claims its own presence, with its own existential content and proudly flaunted diversity. What, then, is "human," if this adjective can no longer be solely attributed to the expression of *Homo sapiens*? The answer is not straightforward, but it helps us understand why the end of ontological anthropocentrism is characterised by a reevaluation of the overlap between the terms "human" and "manifestation of *Homo sapiens*." We may say that "human" is everything we can access and recognize ourselves in. It is the ability to listen to and relate to Roy Betti's story in *Blade Runner* before his time is up. It is an eco-imagination that makes us feel more connected to *all forms of life*. It is the empathy that allows us to resonate even with artificial intelligence. The cyborg is not a superhero; it is human because it embodies the everyday, inhabiting situations that mirror our own anxieties. Its strength also lies in its vulnerability, as seen in William Gibson's *Johnny Mnemonic* (1981). The 1980s saw increasing suggestions of a metamorphic future of the body, and the artistic and narrative worlds of the time bear witness to this.

Let us take a step back in time to understand how information technology had already set the stage for this quantum leap in the way humans are understood through technology. A prime example is Miller, Galanter, and Pribram's book titled *Plans and the Structure of Behavior* (2013), which profoundly changed the way we consider operational processes, building a functional isomorphism between the elaborative processes of computers and the strategic-tactical processes of the psyche. Cognitive psychology, which at the time was undermining the primacy of behaviourism, incurred a significant debt from the functions of data processing. The assimilation is such that, in its most radical formulations, it posits the identity between brain and hardware, on the one hand, and between mind and software, on the other. The explanation of behaviour seems to lose its drive-related motives or the tropistic notion of reaction to stimuli, sanctioning the absolute prevalence of the elaborative and representational criterion. This interpretation of convergence between brain and computer supports not only modelling in psychology but also the increasingly widespread idea that the organic medium can interact with the cybernetic one (Johnson-Laird, 1988).

The immediate focus of interest concerns the potential for expanding perceptual processes, ranging from cochlear implants to retinal prostheses. Initially, bionics aimed to enhance or create exoskeletons, as seen in Japanese manga (Shirow, 1989), but this is merely an illusion. The cyborg is not just a superior human being, but rather something else altogether – the product of the infectious-infiltrative process of the technological apparatus. Around this mutation, two horizons of thought emerged: i) *transhumanism*, which believes that a mutation of the species is underway, a genuine ontological emergence that can elevate humanity to another existential dimension; and ii) *posthumanism*, which, conversely, regards the cyborg as an opportunity to re-examine the humanist paradigm of human self-sufficiency and cultivate a new philosophy that regards predicates as the product of hybridisation. The latter perspective does not discount the prominent role that the latest technologies have played in exacerbating the metamorphosis of humanity, but it links this process to an ontological condition that has characterised human beings since their earliest cultural expressions. For posthumanism, the cyborg is not only a blueprint for conceptualising the humanity to come, but also a means of understanding the present human condition in its dependence and fragility.

Something significant is undeniably happening. The cyborg appears as a visionary, striving for a progressive detachment from the confirming logic of Gestalts. While the world is mediated through sensory receptors throughout the body, the cyborg seeks to circumvent them. Instead of producing the expected panorama, its world will be one of shifting mediations, far removed from the traditional senses that were preset by phylogeny (Hughes, 2001). The traditional role of the senses was to filter reality and present it to the mind. Now the aspiration for a direct dialogue between technology and brain paves the way to a journey into the unknown, rather than just enhancing existing abilities. If sensory inputs, with their inherent subjectivity, used to reveal a reality that was consistent with the mind's preconceptions, the introduction of sensory chips changes things dramatically. By grafting a sensory chip, the cyborg can assume different receptivities, such as those of other animals, sensitive to sound or electromagnetic frequencies that are inaccessible to us. By altering our evolutionary access to the world, we disrupt the mental mechanisms that organise our sensory experiences. This can cause upheaval in the mental mechanisms of organising referents. So this is not simply about enhancing our existing senses, but about creating a new, psychedelic condition that we have yet to fully comprehend. It is now clear that this dream is slowly taking shape with the increasing development of digital technologies, where psychedelia is becoming a way of life (Leary et al., 1964).

Still focused on phenomenology, perception is at the centre of this anthropological metamorphosis. It is not merely a sensory mutation, but an unprecedented access point to a new existential dimension. As Maurice Merleau-Ponty argues, we see much more than what the eyes capture, and perception is not only a receptive process: it is also predicative. The mind is not a mere container for sensory phenomena, but it is responsible for organising the content. The senses serve as tools that the mind employs to attribute phenomenological qualities to appearances (Merleau-Ponty, 1964). When the brain communicates directly with technological devices to access the world, the mind acquires new tools to construct reality. However, if sensory tools are like the words we use to converse with others, encountering new sensory experiences can leave us speechless. Sensory organs and mental a-prioris are interdependent and provide a foundation of stability for our experiences. Without this stability, one risks becoming lost in a hyperstimulative fog.

It should come as no surprise, therefore, that the genesis of the cyborg and subsequent projections within virtual realities are accompanied by a profound transformation, a true existential mutation of the individual. The error lies in believing that bypassing the sensory transducer provides access to reality only in extensional terms: that is, as additional information. If, however, we understand perception in a predicative sense, as the construction of new planes of reality, it is evident that change would affect far more than the objective qualities of the received phenomenon. Indeed, it would be a matter of *inaugurating a new phenomenology of reality* (Henry, 1963). This should be the real focus of the discussion on grafting information technologies, starting from some considerations that we can already make by observing the cognitive transformations induced by digital technologies. The fact that an adolescent who spends eight hours a day immersed in digital media exhibits a different configuration of attention, concentration, and stress mechanisms already provides us with useful insights into this transformation (Spitzer, 2012). The cyborg is exemplary in showing us how a different dialogue with the world has the power to change the entire existential dimension. This is the reason why we should recognize its heuristic value in decrypting the future.

3.3 The body under accusation and the body under pressure

The phenomenology of the cyborg, if we momentarily suspend any judgement of value regarding the most extreme design expressions, should be understood as a heightened awareness of the multifaceted significance of human expression. Indeed, the cyborg represents an acknowledgment of the impossibility of human beings to strip down to their naked body, as the Vitruvian Man claimed to do. Every human predicate must be seen as emerging from a process of hybridisation. In the cyborg, the body is inextricably and indivisibly intertwined with technological apparatuses that cannot be regarded as mere additions or supplements but are rather integral parts of the somatic dimension itself. Technological prostheses engage with the body, but they do not merely enhance or extend its qualities in a quantitative sense. Rather, they modify the body, altering its coordinates and introducing new ones. Even if we could hypothetically strip down the human being – an assumption that disregards the internal and organisational rootedness of the body's morpho-functional apparatuses – we would lose the recognizability of the human. It would be like erasing the physiognomy of our face. Indeed, the compensatory conception is fundamentally flawed, as the device does not merely construct a functional barycentre for the body, but projects the somatic dimension onto a new plane of sensibility, agency, and even existential (perspectival and teleological) significance.

While the complementary view may offer greater stability and protection, the hybridising principle of the cyborg exposes the body to a battery of stress-tests. Indeed, it is evident that any shift in perspective entails problematic aspects. The body is compelled to maintain itself in a state of non-equilibrium; it has heightened its *heteronomic character*, which means its reliance on external formatting, resulting in a loss of autonomy (Marchesini, 2017). Additionally, it is increasingly challenged by an environment that is less and less consistent with its phylogenetic a-priori, leaving it more precarious, fluid, and exposed (Plessner, 1928). What humanists attributed to man's original state, the forgetfulness of Epimetheus, is in fact the consequence of the technopoietic progression of human beings who have made their somatic dimension less stable and static by utilising increasingly complex and pervasive technologies. Thus, the cyborg serves as an interpretive model that enables us to retrospectively examine anthropology while prefiguring plausible scenarios for the future. It is important to note that the technopoietic trend has not grown linearly but exponentially. This suggests a likely remarkable leap in the 21st century (barring any unforeseen incidents such as an ecological catastrophe), which some authors refer to as the *singularity* (Vinge, 1993).

The rapid advancement in technomediation raises numerous questions, including the problem of artificial intelligences becoming increasingly autonomous and influential in human life. Furthermore, even with the current state of digital technology, which assumes the existence of metaworlds capable of replacing concrete experience, it is evident that the body is undergoing a period of significant compromise. With this in mind, it's useful to distinguish between the problems that involve distortions in the interpretation of the body in the technomediated context – the "inadequate body" – and the critical issues that concern somatic suffering, or the "suffering body" due to invasive interference by technology. We can differentiate these two orders of problematicity as follows: 1) the *body under accusation* for its inability to adapt to the rapid technological development, opposing and resisting change due to its organic constraints; 2) *the body under pressure* from technological perfusion, especially from the digital revolution and its totalising effect on human life. I believe it is mistaken to view the cyborg exclusively through its heuristic traits or with a blindly enthusiastic approach, because alongside its undeniably positive aspects – like showing us the dependencies of human beings on the world and recognising the close relationship between technology and the implementation of needs – it also brings with it distortions and critical issues when it becomes a negation of the body.

With regard to the first point (the body under accusation), it is evident that the accelerating pace of recent decades has widened the perceived gap between the potential of the body and that of technology. Despite the misunderstandings of the somatic dimension still prevalent in the humanist legacy, it is undeniable that the body, precisely because it is adaptive, presents constraints to technological reconfiguration. Technological development operates within historical times, which are now accelerating beyond generational trends. In contrast, somatic organisation, while flexible in ontogenetic terms, still has a conformation that reflects biological timescales, spanning millions of years (Yan et al., 2020). Until the 19th century, technological revolutions occurred at a pace of at least a hundred years or so, but by the next century they were measured in decades. Today, the pace of innovation lies within a few years (Xie et al. 2019). As a result, the body has no time to adapt to a given technology before it becomes outdated and is replaced by a new one that maintains little continuity with the previous one, relying on different technical principles. Thus, the body undergoes constant adaptive stress, struggling to keep up with an ever-changing technosphere, and scrambling to stay in the same adaptive spot.

The cyborg model is at risk of being too heavily focused on the technological component, which can distort the understanding of hybridisation itself (Gawdat, 2021). In fact, this can lead to a shift from an awareness of the conjugative potentialities of the body – such as its plasticity – to a *renewed perception of inadequacy*, only in the opposite sense compared to the classical humanist view. Indeed, according to the Gehlenian perspective, the human body is incomplete because it is incapable of acting in the world and requires the technological crutch. Instead, the new sense of lack underlying the cyborg ideology refers to the body's inability to keep up with the potentialities offered by technology itself. An analysis that aims to highlight the merits and demerits of new ways of interpreting the body-*techne* relationship must also address the present or potential drifts of this ideologised image of the contemporary or future cyborg. The following aspects, in my view, are the most problematic:

1. The tendency to view the body as an obsolete entity, too constrained by phylogenetic reasons, and therefore not very malleable compared to the dynamism of technological development.

2. The disregard for the reasons of the body and the desire to subject it to performative regimes that can negatively impact well-being.

3. The absurd idea of completely getting rid of the somatic dimension.

If the Vitruvian Man claimed to recognise himself by stripping off his clothes, the cyborg risks claiming to *find himself by depriving himself of the body altogether*. While the third point may sound paradoxical, neglecting the body in pursuit of the illusion of techno-transcendence is a real danger (Kurzweil, 1999).

Therefore, analysing the possible consequences of cyborg ideology calls for avoiding simplistic fascinations and adopting a critical attitude. We should reject the essentialism of Vitruvianism while being fully aware that every paradigm shift leaves open problematic issues that we must address. Acknowledging that the body is not ontologically self-sufficient and cannot explain its predicates without external factors does not entail denying its precise relevance or turning it into a fluid entity. The body has its own form and cannot take on any externally imposed shape or be replaced by a technological apparatus. The correct way to interpret the cyborg is to recognize the value of the relational dimension of the body, without denying its inherence or advocating for its full surrogacy (Leonhard, 2016). Unfortunately, the risk of shifting to the opposite bank is not remote, as demonstrated by transhumanist philosophy. The error lies in ignoring, denying, or neglecting the *specific* attributes of the somatic dimension that cannot be assimilated, superimposed, or substituted by technological devices (Fukuyama, 2003). We must differentiate between considering the tool as a prosthesis that is incorporated and considering the device as equivalent to an organ.

The mechanic model can be helpful in explaining certain body functions because this requires a linear cause-and-effect relationship, a precise temporal sequence, controllable explanatory mechanisms, a priori choices of scale, and the exclusion of complementarity, recursiveness, and chaoticity by labelling them as background noise. This allows us to extract a clear causal mechanism from the process, which can even be replicated in a medium. Machines have been epistemologically valuable because they allow for the simulation of certain processes by reproducing them under controlled conditions. However, using a model for epistemological purposes is one thing; positing a complete overlap between the machine prototype and the phenomenon itself is quite another. When dealing with complex processes – especially the phenomenon of life, which is often irreducible to a mechanistic model – we must be careful not to confuse epistemology with ontology. While a computer can serve as a useful metaphor for understanding some of the brain's processing mechanisms, it is a mistake to view the brain as nothing more than hardware running within an operating system made of flesh (McCarthy et al. 2006).

The cognitivist idea, which is based on the concept of the internal processor, interprets the mind solely in terms of processing dynamics. This approach led to the conception of identity as a collection of functions and contents present in a brain that is thought of exclusively as a computational entity (Von Neumann, 1958). This quickly led to the idea of an assimilative or isomorphic relationship between the brain and computer, paving the way for the dream of mind-downloading and a new form of immortality (Bamford, 2012). If what defines us can be reduced to a mere set of data running on a biological medium, and if its existential manifestation is realised through algorithmic dynamics, the transferability of subjectivity from the brain to another medium becomes plausible (Wiley, 2014). When even the most intimate aspect of subjectivity – the mind with its biographical, cognitive, and affective structure – can be transferred from the biological medium to the technological one, it becomes evident that the body is no longer a necessary anchor for subjectivity (Kurzweil, 2012). Here

I do not want to consider either the presupposed isomorphic conception or the feasibility of the process itself at this moment, but I take their formal correctness for granted. Still, it is clear that we are no longer dealing with the phenomenology of the cyborg, but rather with another entity. By removing the body's *specific* characteristics, we enter a realm that is no longer hybrid, but entirely mechanical.

This paradigmatic metamorphosis is not necessarily implicit in the cyborg model, but it may result from an overemphasis on the value of the hybridising medium at the expense of the body and the organic dimension. The cyborg thus presents itself as a transitional figure useful for ferrying the human being into a no-longer-human condition. It is important to understand how this revolution in thinking that made the body a surrogate entity through technology came about. Some answers come from the Cartesian dictate itself, except that the French philosopher was careful not to reduce the mind to *res extensa*. Since the 1960s, instead, computer science has ushered in a new cultural climate marked by a fascination with a machine capable of performing tasks hitherto strictly within the domain of the human brain – hence a progressive collaboration between psychological inquiry and technological simulation (Minsky and Seymour, 1972). The computer provides new models for entering the black box of the mind and understanding cognitive processes. The side-by-side comparison of brain and computer becomes even clearer with the emergence of increasingly high-performance software, capable of winning against human beings even in activities deemed to pertain to our intelligence. With the development of deep learning at the threshold of the 21st century, we have come to the self-programming of artificial intelligences, and at this point, the hiatus seems to be finally bridged (Deng and Yu, 2014).

In the 1990s, neurobiology also made a significant contribution towards the convergence, if not overlap, between the processing conception of computer models and the physiology of the brain. One example is the connectionist model, which, unlike the classical stimulus-response associations of behaviourism, demonstrates how output can differ as the internal patterns of the system vary, where these functional states are obtainable through different binding configurations of various points in a network (Rumelhart and McClelland, 1986). The assimilation of brain and computer soon appeared to be something that goes beyond a simple epistemic stratagem. By arranging neural networks capable of connected activation state functions, cognitive science will be able to show significant procedural similarities with the brain connectome. On the other hand, neuron activation and its discharge function are reminiscent of binary code, whereby synaptic structure can be translated into a sequence of 0/1 elements, and mental contents can be rewritten into a wiring code. At this point, the computer increasingly appears to no longer be an epistemological metaphor, and is described as a kind of biological hardware (Kerskens and Perez, 2022). In the same vein, if one can identify the neural configuration of a memory, it becomes possible to artificially insert a memory trace without the need for a specific experience to generate it.

This cultural climate, which characterised the last decade of the 20th century, is increasingly inclined to consider the neurobiological apparatus as nothing more than an internal processor, realising Philip Dick's visions of grafted memories and identity reconfigurations. As a transporter of data, the cyborg can load information into its brain hardware, which can then be transferred to another medium. This conception of the interface between brain and computer gives rise to fantasies about reconfiguring the psyche and suggests the idea of a communication with technology that is no longer mediated. However, this vision of the body, which loses its lived experience to become exclusively the substratum of a *res informatica* devoid of biological identity, already heralds the annihilation of the somatic dimension itself. The container body is amorphous, lacking its own implicit intentionality, and can therefore be abandoned at the right technological opportunity. One cannot understand the cultural climate of the early 21st century without referring to a generic *pygmalion tendency* that absorbs human beings, leading them to consider the organic dimension as obsolete due to its strong ties to phylogenetic adaptation (Hilton, 1993).

In cyborg ideology, the brain is viewed as complex hardware, but not necessarily effective or efficient when compared to the possibilities of quantum computation, for instance. This is not a simple assimilation of the brain to the computer, but a declaration of the *inadequacy of the body* in the face of the potential of new technologies. The somatic dimension is being criticised by transhumanist theorists who aspire to a new form of technocentric transcendence based on emancipation from the body. It is important to distinguish between: 1) the classical phenomenological conception of the cyborg, which is seen as the bearer of its own dialogical content and considers the body as a chiasmatic point of relation with and through technomediation, and 2) the transitional figure of the cyborg as a moment of experimentation towards a progressive substitution of the body, which simply serves as a bridge to a condition that can only be described as no-longer-human (Goertzel, 2007). In my opinion, this is a dangerous drift that, while different, is similar to the shift from the hermetic humanism of

15th-century philosophers to the post-Cartesian anthropocentric humanism of emancipation from nature.

The tendency to view the cyborg as a transitional figure has been reinforced in the first few decades of the 21st century by the development of self-programming artificial intelligences and advancements in neurobiology that reveal the mechanisms of the psyche. Neuroscientific studies of different brain areas involved in various behaviours and neuromodulation pathways that stimulate certain psychological conditions seem to support the notion of technological control of the psyche (Baker, 2011), a first step toward techno-substitution. Furthermore, new practices of neurostimulation through electrodes confirm the possibility of grafting chips capable of direct dialogue with the brain. It is now possible to induce emotions using activators or reuptake inhibitors of certain modulators or to induce specific body movements by electrically stimulating certain brain areas. As technology becomes increasingly invasive, the body becomes more transparent and permeable, thanks to the discoveries of late-millennium biology (Eggebrecht et al., 2014).

As soon as the body is controlled by technology, we witness a paradigm shift: the body becomes an instrument of the device, instead of the other way around. So, just as any tool can be replaced, discarded, or improved upon, the instrumentalised body is susceptible to being perceived as obsolete. This epochal transformation profoundly alters the phenomenology of the cyborg, transforming it into a transient entity awaiting realisation through the abandonment of the organic dimension, which is now considered a redundant vestige (Dormehl, 2014). Consequently, it is easy to see this new *transitional cyborg* as a sort of ferryman, a modern-day Charon whose role is to transport humanity to a different existential shore. Its body, studded with functional access points and gateways to a diverse range of technological connectivities, is no longer an expression of the implicit relational condition of the somatic structure but is instead viewed as an animal heritage to be detached from. If the phenomenology of the cyborg as an expression of hybridisation challenged the ontocentric view of the Vitruvian Man, the new transhumanist ideology, by interpreting the cyborg as a transitional figure, renews the same aspiration of transcendence through its transfer into a silicon medium (Ponti, 2016).

The fluid and magmatic nature of technology often leads to an accusation against the body, also thanks to the traditional naturophobia of Western culture (Keith, 1983). What are the accusations against the body? First of all, its fragility, ageing, and transience, which stand in contrast to technology's ability to seemingly defy the wear and tear of time. But there is another aspect that should not be underestimated: the uneasy proximity of the somatic dimension to the animal world, made even more problematic after the Darwinian revolution. Technology becomes a means of permanently escaping the stigma of common ancestry: emancipation from the somatic dimension coincides with the abandonment of animality. Finally, the body's physiological needs seem to weigh it down and oppose transcendence, the dream of purity and distance from the telluric underlying the humanist imagination (O'Connel, 2018). In contrast, technology is portrayed as free from contamination. This transhuman cyborg aspires to go beyond its animal nature: by shedding its nipples and sexual characteristics, as well as any vestiges that no longer conform to its *Dasein*, it prepares for its final metamorphosis (Horton and Priest, 2018).

There is, however, an alternative portrayal of the cyborg that dominates the dystopian scripts of the end of the millennium, appearing in science fiction that explores the post-apocalyptic world. This portrayal depicts a body that is suffering due to the altered environment and the scarcity of untainted resources. The image is that of a somatic dimension that is no longer adaptive to changing ecological conditions, a body that is once again ancestral, like a resonance of a past to which there is no return. The cyborg moves in a constantly threatening underground, equipped only with leftover technologies, creatively piecing together devices repurposed and plundered from the dumps of the past. The needs of the body do not aid the cyborg, on the contrary, they make survival even more difficult and accentuate its adaptive distance from the machines, which, equipped with artificial intelligence, armoured in metal, and devoid of metabolism, are perfectly at home in that dystopian scenario. The body suffers in its confrontation with a threatening technology, in a world that has irrevocably disrupted its living conditions. The cyborg, then, is transitional as it dons the shoes of the refugee and the stowaway who survives in the folds of the world, constantly on the run in the face of the powers that be. The body no longer belongs to the cyborg, studded as it is with multiple switches for new networked devices to access. Rather, the body leaves the cyborg at the mercy of the world's interference.

As can be seen, the phenomenology of the cyborg is far more complex than the vision of it presented in Paul Verhoeven's *RoboCop* (1987), where the cyborg is portrayed as a superhero who eradicates crime with mechanical prosthetics, titanium armour, and a brain integrated with a computer system. In reality, the cyborg is exposed to the world and is a plastic expression of fragility. Even in *RoboCop*, there is a foreshadowing of ambivalence as the transformed agent Murphy seeks rootedness in his former life and ultimately reclaims his identity. This raises the question of whether the cyborg is a human who has enhanced their performance or if the body inevitably ends up serving the artificial intelligence to which it is connected. It is clear that if artificial intelligence surpasses human intelligence, as is already the case in many fields, the latter option will be far more likely. The cyborg, therefore, proves to be a far more problematic entity than is usually claimed by anthropocentric hubris. This model reveals the unpredictable character of any infiltrative technology, which enhances certain functions but also paves the way to conditions that may only be recognised in retrospect. Technologies reposition human ontology on perspectives beyond human imagination, particularly in the case of information and digital technologies.

Let us now consider the second point: the body under pressure. I will only provide a few examples here, before examining the physiological alterations caused by technology in detail in the following chapters. We can identify several ways in which technology puts pressure on the body, forcing it into compensatory actions that often lead to impairment. For instance, (i) technological support requires the body to modify its performance by pushing its ability to readjust; (ii) technomediation increases the overall level of performance, but also leads the body to complementary overwork; (iii) each technology alters the body's balance by causing wear and tear in certain areas and deficits in others, resulting in a system imbalance that makes other functions difficult; (iv) the presence of high-performance devices raises the level of social competition, demanding that individuals increase their resources to keep up; (v) the set of technological supports changes people's living environment, creating contexts and situations that can incentivize certain pathologies; (vi) the presence of certain technologies modifies relationships between people, changing individual aspirations and investing existential meanings, thus globally affecting the efforts required to make sense of one's life.

Man invents a device, but soon this redefines the coordinates of humanity itself, bringing its existential condition into a fundamentally different epistemology, ontology, and eschatology. By examining the cyborg model, we can grasp not only how the imagination responds to new technological possibilities, but also the profound implications of the integration of body and technology, which cannot be confined to a purely performative realm. The term "tool" is misleading, especially in the context of digital technologies or biosynthesis, as it conveys a passive, subordinate, and non-recursive role for technology in relation to the human condition, under our complete control. Indeed, the cyborg challenges this notion, revealing that the body is not empowered by technology, but rather transformed in a new organisational and functional framework. The body is not master of technology, simply gaining a performance-based return when utilising a device. In reality, the body is the first to be involved – and sometimes even questioned – when human existence becomes intertwined with a new technology. The body is subjected to stress, dissection, reconfiguration, adaptation to new functions, dependence and, at times, even intoxication by what we mistakenly continue to perceive as mere tools (Cregan-Reid, 2018).

Today, we are aware that new digital technologies can lead to increased arousal, standardised performance, deficits in attention and concentration, heightened anxious and aggressive responses, sleep disorders, decreased memory skills, greater (dis-)stress, dopaminergic dependence, and a heightened feeling of vulnerability (Dainoff, 2007). This transformation has only emerged in the past decade, yet scientific evidence already points to its impairing effects, particularly in the psychological, affective, and cognitive dimensions of individuals. The most concerning findings are observed in the so-called millennial generation, those born after the rise of digital technologies and social media. Continuous use of these technologies has consistently affected the cognitive performance of children, with discouraging outcomes in areas such as concentration, memory, text interpretation, and information retrieval (Ophir et al. 2009). However, alongside these deficiencies, there is an increase in other cognitive qualities, such as response speed, focus, salience extraction, and web navigation. This underscores the inherent evolutionary differential associated with digital engagement.

The cardiovascular system is perhaps one of the areas most impacted by the new habits introduced by information technology, albeit indirectly: through a significant decrease in physical activity (Warren et al., 2010). This is evident in the increased number of cardiovascular diseases, which currently account for approximately 35 percent of deaths. The consequences include elevated blood pressure and higher levels of saturated fats in the blood, both of which can be attributed to stress mechanisms. The body is therefore strained in order to cope with the performance levels imposed by these new habits, to compensate for the disruptions caused, and to restore metabolic equilibrium by addressing excess free radicals and acidity gradients, among other factors. Considering the device as a mere facilitator of functions is a gross oversight, as the body is always required to define an overall positional barycentre, or balance point, that aligns all the functions of the organism. Furthermore, the resources available to mitigate system dysfunctionality are limited, particularly given that many of the imbalances mentioned above can trigger synergistic effects or chain reactions with self-perpetuating consequences.

The impact of technology on the body is not limited to its direct effects, but also stems from the heightened performance that it requires. Each device raises the bar in terms of what is considered acceptable performance: from keeping up with bureaucratic and organisational mechanisms of society, to competing with rivals or meeting user demands, to staying consistently informed in the era of globalisation with increased expectations of efficacy and efficiency. This has led to a significant rise in the use of anti-anxiety drugs, with annual growth trends approaching 10 percent over the past decade (Garakani et al., 2020). The influence of technology is not confined to the individual in an unchanged context, but rather results in a transformation of the world itself, including the rules of the game and the goals required. It is in this context that its repercussions must be considered. Technological innovation does not simply benefit the individual, but also imposes new requirements for citizenship. Each technology demands that individuals acquire the necessary resources to obtain it, so as to participate on equal terms in the civic realm (Castells, 1999). And even aside from the phenomenon of the technological divide, it is evident that every innovation demands effort from the individual, whether it be financial or otherwise, in order to fully participate in the broader community (Turkle, 2011).

The physiology of the cyborg has undergone a metamorphosis to cope with the demands of technomediation, as evident from the stress and anxiety introduced by the digital revolution into our lives, alongside its undeniable benefits. Viewing devices solely in terms of empowerment is therefore inaccurate for two reasons: i) because what is actually produced is not just empowerment, but a transformation in the overall experience of the individual, starting from the body; ii) because often the body itself is affected by the impact of new technology, with effects that become apparent only after some time, when pathologies start to manifest. For instance, consider the damage caused by pesticides, plastics, electromagnetic waves, radioactivity, and the use of auxins in animal husbandry. The body is increasingly under pressure by what should be seen as genuine *techno-coercion*, resulting in consequences that people bear the brunt of. It is not about rejecting technological development, but about critically understanding its impact on our bodies in order to identify the best solutions for promoting well-being, recognising that the plasticity of the body is not synonymous with fluidity, and that our existential dimension is grounded in being a body, albeit one that is open to hybridisation with the world.

CHAPTER 4

THE INSUFFICIENCY OF THE TOOL MODEL

Premise

To better understand the impact of technology on the body, it is important to move away from the traditional model that views technology as a mere instrument (Gibson and Ingold, 1993). This framework, in fact, sees technology as a passive object that performs tasks at the individual's command, acting only on the external world – such as a hammer hitting a nail – and under the human being's full control. This view is rooted in an anthropocentric concept of technical action that assumes that human ontology, including the body, is impervious to external influences. It sees technology as a unidirectional intermediary between humans and the world (Leroi-Gourham, 1945). The instrument is seen as incapable of bringing about somatic transformations or enabling modification processes of the human being. This concept of technology as a tool encapsulates values and principles that have characterised the modern age since the Renaissance (Scheler, 1955).

Now, questioning the concept of tool does not mean denying the instrumental use of devices, but rather trying to understand the implications that this model entails and testing its consistency (Kortland and Kooij, 1963). The concept of tool, in fact, is not neutral and does not merely define a specific intended use or the device's capacity to enable certain operations (Beck, 1980). It also carries other connotations such as: (i) being external to the body, unrelated to somatic physiology, and not affecting the purity of the organism; (ii) being subject to the total ownership of the user, who has full control over its variations of use; (iii) being designed according to a predetermined performative canon, created for a specific purpose; (iv) being ergonomic, tailored to fit the human body and enhance its capabilities; (v) being used to act on the world in a unidirectional way, based on a particular vision of human agency that modifies reality without being modified by it. These implicit characteristics, inherent in the concept of tool, clash with the development of new technologies, which have made their inadequacy apparent (Dyson, 1997).

Referring to a device as a mere tool implies accepting a cultural paradigm that views the human being as impervious to any possible feedback phenomenon, whether caused by the device or by reverse directionality from the world to the human. Furthermore, the passivity attributed to the tool not only renders it incapable of affecting the body, but also gives the human being who uses it the power to define its type of use from its design stage. These assumptions are severely challenged by the technologies developed in the 20th century, which emphasise characteristics that have actually been present since the earliest human technical expressions (Habermas, 2001). Even a simple chopper, for example, reshapes somatic structure, lends itself to unforeseen uses, influences the relationship with the outside world, and forces the body to acquire certain performances. As devices become more complex, their instrumental connotations tend to blur, revealing numerous backlash phenomena, such as the somatisation of technology. Therefore, questioning the concept of tool entails considering the modulation that technology exerts on the physiology of the organism.

Conversely, if we were to consider the device as a somatic agent, it would possess the following qualities: (i) *Infiltrativity*, as it is always somatised, even if it appears to be located outside the body. This means that all technologies, no matter how simple, produce hybridisation and are embodied. (ii) *Proactivity*, meaning it is never completely in human hands, as mere possession changes the range of possibilities and influences all decision-making processes. This is especially true for current artificial intelligences, which are by definition beyond human control. (iii) *Co-optability*, as it invents new canons of application that go beyond what is imagined by humans. Information technologies, for instance, inaugurated unpredictable operational spaces, creating new existential territories that modify the human teleological dimensions. (iv) *Normativity*, meaning it does not adapt completely to the body, as the flesh exhibits greater pliability, shaping itself around the technology. This is also due to the metamorphic impact that the device produces on physiology. (v) *Bidirectionality*, as the technological intermediary increases the influence of the world on the body, broadening the horizon of interactivity with external reality.

It is now clear that technology has always modified the somatic organisation of humans instead of passively fulfilling the desires of human beings (Cregan-Reid, 2018). We can identify three main orders of bodily metamorphoses operated by devices: (1) *metabolic*, as the device changes the physiological rhythm of the body, affecting all functional cycles and their regulators, such as hormones, cytokines, and neurotransmitters; (2) *ontogenetic*, as it produces a growth differential in different tissues and often modifies the relationships between various parts of the organism; and (3) *phylogenetic*, since, according to niche theory, every transformation of the context changes selective pressures and thus influences the evolutionary trajectory of a species. The most notable phenomenon is phenotypic induction, given that the shape of an organism is one of the possible variations of the genotype, induced by environmental variables acting in ontogeny (Gilbert and Epel, 2015). The technosphere is a developmental environment that intervenes in the inherent polyphenism of the human genome, which is the ability of our DNA to produce different phenotypes depending on the environmental variables present.

Using new models, we can better understand the concept of technophysiology as a biological intervention by technology on the body, which significantly impacts metabolic profiles, modulatory arrangements, cell growth, embryonic processes, functional triggers, and fitness mechanisms. In other words, the device behaves toward the body like a hormone, a neurotransmitter, an immune mediator, a virus, a symbiotic microorganism, a growth substrate, a selective pressure element, a vitamin, or a free radical. Examining these patterns helps us grasp the infiltrativity of technology, showing why its influence cannot be reduced to a generic effect of enhancing or extending human performance. The device transforms the body rather than just serving as a surrogate, because the action of morpho-functional modification involves a wide variety of interventions (Clynes and Kline, 1960).

Discussing models of device representation is a prerequisite for a new *philosophy of technology* (Feenberg, 1999) that can address the challenges posed by information technology, nanotechnology, and biosynthesis. Devices used by humans have always modified their creators – there is nothing new about that – but now physiological recursiveness and somatic metamorphosis have taken on a scope that can no longer be ignored. The new technologies demand, therefore, a much deeper analysis, capable of embracing all their repercussions on the human condition, beyond performative horizons alone. Performance may fascinate or appal – think, for example, of the different attitudes still held towards the evolution of artificial intelligences capable of outperforming human capabilities – but it remains confined to the obsolete idea of the externality of technology (Mitchell, 2019). Any reflection is misleading if developed within the conceptual perimeter of the tool: it risks making us believe that the problem of technology can be confined exclusively to the realm of management. In other words, we continue to make two incorrect assumptions: 1) the immutability of human predicates, i.e., the idea that the essential core of humanity is beyond the influence of technology; 2) the passivity of the device, i.e., its ancillary and essentially inert role as an entity whose use is fully under human control and which does not challenge the existential coordinates of human beings (Rümelin and Weidenfeld, 2018).

As mentioned earlier, this approach is inadequate even for old technologies, as each device is embodied and produces somatic transformations that go beyond mere compensation or expansion of operation (Brey, 2000). However, the effects of these transformations were less explicit, more subterranean, and more manageable. With the new technological revolutions, this interpretation is no longer viable. The *five qualities of somatic agency* in the device – infiltrativity, proactivity, co-optability, normativity, bidirectionality – have become evident, rendering the concept of tool obsolete. We are now faced with a real decline of the conception of technology as offering a mere dressing of the body and passive servitude to human purposes (Kelly, 2010). This eclipse deprives us of a horizon of meaning that constituted the fundamental reference of technomediation and anthropology in the modern age. A new era is upon us, ushered in by new devices that differ significantly from those we are familiar with, which is the product of a shift in interpretation of technology that unsettles us by removing the certainties of domination that characterised human Prometheanism. Orphaned of such a reassuring paradigm, we find ourselves moving on treacherous and ambivalent ground, without the comfort of being able to retrace our steps (Hockfield, 2019).

The reason why we keep referring to a generic *tool*, with all the meanings this implies, is due to our cultural unpreparedness to understand the phenomenology of technology during this transitional period (Borgmann, 1984). Our relationship with machines has evolved rapidly over the past two hundred years, and we have been somewhat caught off guard by their potentialities and their effects on our lives. We have been fascinated by their performance, like a child at an amusement park, assuming that nothing would change except for our increased power over the world (Norman, 1994). Trapped in the dichotomy between wonder and horror, techno-enthusiasm and neo-Luddism, we have dreamt of a transhumanity in the digital realm or wished for a utopian return to nature, *à la* Walden, without realising that the very terms of the debate have shifted, and our environment has undergone an irreversible metamorphosis. To rethink the concept of technology, we must accept the challenge of an adventure in the unknown, which requires new maps – that is, new conceptual tools able to illuminate phenomena that have been forcibly kept in the shadows by the prevailing paradigm but are crucial to deciphering the future (Harari, 2011).

A timely reinterpretation of our assessment models should be prompted by the greater infiltrative and impactful gradient of new technologies on the body. The information technology revolution has radically changed human life by creating a global communication space that is now impossible to escape (Bostrom, 2018). The body has become a gateway for a flow of information that goes beyond the organisational possibility of the human being. Therefore, the problem is no longer how to acquire information, but how to manage it. The temporal dimension of experience also changed, as all processing became increasingly accelerated and less consistent with the cognitive mechanisms of our brains. This led to a constant condition of stupor that limits the processes of elaboration and deeper analysis (Demichelis, 2022). While the advent of the Web has provided undoubted opportunities, it has also forced us to gloss over events, initiating an attitude of detachment towards information, which thus loses adherence to our biographical substratum. This overflowing stream desensitises our capacity to acquire information, turning everything into an exciting spectacle leaving nothing but scattered mnemonic traces. We can clearly see the subject's loss of ownership in the information process due to an excess of availability – it's as if, in front of a sumptuous lunch, we were gripped by nausea or bulimia.

The emergence of the digital realm has led individuals to become fully immersed in a distorted sense of reality, as they believe that online is where everything happens (Searle, 1995). The physical world, including the home, office, metro, or park, is relegated to a secondary role, becoming nothing more than a background for events that take place online. In this realm, time appears to stand still, in contrast to the frenetic pace of events on the web, which exerts an irresistible pull with its attractive rhythm (Illich, 1973). The digital world, with its endless opportunities for consumption, is a labyrinth that entraps individuals in a regressive state, reminiscent of infancy. The constant need for connection can stimulate an oral fixation towards news or even foetal passivity, reminiscent of the placental flow. This experience is all-consuming and often feeds an insatiable narcissism in individuals, who are unable to escape the lure of continuous connection (Floridi, 2014). So can we still interpret these technologies with the traditional model? At first glance, it seems possible, as *we* are the ones accessing Google, Instagram, Spotify, Netflix, WhatsApp, or Facebook. But in reality, these technologies are far more complex and ambivalent (Lawson, 2018). They not only serve us but also question us, profile us, and propose to us what they think we'll find desirable, feeding our insatiable narcissism (Frankel and Krebs, 2021). They have radically altered fundamental aspects of society, such as democratic participation, opinion building, social conviviality, leisure time, and ultimately, the values and priorities on which we base our existence (Morozov, 2011).

The concept of technology as a mere tool obstructs our understanding of its infiltrative and pervasive potential, giving us the illusion of full control over, and detachment from, its impact (Bachimont, 2010). The urgent need to revise our models of interpretation arises from the superior ability of new technologies to affect our physiology. In fact, it is not enough to update our operational systems, but we have to update our conceptual framework as well. This need is further emphasised by (i) the accelerating trend of technological innovation, (ii) the converging trend of new technologies, and (iii) the perfusive and connective nature of these technologies. Regarding the first point, it is clear that we are facing an exponential growth of technological advancement that will only increase in the near future. As for the second point, the synergy between information technology and communication, in addition to yielding mutual enhancement between various technical fields, has produced emergent effects that are difficult to predict (O'Connor and Wong, 2012). This confluence phenomenon is leading us toward a moment of technological singularity, enabled by convergence. Lastly, we must not forget the perfusive propensity of neo-technologies, which are increasingly interconnected and able to control processes that are unmanageable for human beings.

4.1 Technosphere and ontogenetic niche

Every technological device has the power to cause physiological transformations in the body, whether they are minor or major. Investigating all the effects that technology induces on the body would require extensive research that goes well beyond the scope of this essay. However, I think it is still valuable to assess the *ontogenetic modulating impact* that the technosphere (the collective set of technologies and techniques used in a certain era) has on the somatic configuration of individuals (Shilling, 2016). To accomplish this, it is essential to first question the concept of tool, which previously had an exclusively performative relevance and was marked by its passive readiness for use by humans. Instead, we need to redefine the ontogenetic constitution of the somatic dimension and see the body as an active entity that interacts with the world to shape itself, rather than just adapting passively to environmental circumstances. By doing so, we will acquire the necessary conceptual framework to understand the transformation of the body through the symbiosis inaugurated with technology (Warnke, 2017).

To analyse the somatic remodelling repercussions of devices, one must first break free from the tool model and adopt a new conceptual perspective. In the traditional view, humans use devices solely to expand or intensify their operational scope and achieve performances that would otherwise be impossible (Norman, 1994). While not incorrect, this view is partial and misleading. It assumes that devices do not modify human or somatic predicates, acting merely as gloves that safeguard the integrity of the hand. In contrast, recognising that devices also modify the body requires a significant interpretive leap, which is made difficult by the persistence of an anthropocentric conception that presupposes full control over machines and the intangibility of humanity's sacred essence (West-Eberhard, 2003). However, the recursive somatic influence of devices is becoming increasingly relevant and can no longer be neglected in the face of new technologies. Thus, we are witnessing a *metamorphosis of the relationship* between the body and technology, which demands a shift from the anachronistic concept of *techne* as a tool confined to its realm of use to a new model (D'Alessandro and Potestio, 2006).

Therefore, we can assert that technology has a dual function: on the one hand, it serves as a performative entity, which humans can use while also creating new spaces and opportunities for action, often with unpredictable outcomes that challenge the traditional concept of a tool as a mere functional entity under man's full control. On the other hand, technology also influences the internal physiological mechanisms of the body. When we acknowledge this, the image of passivity, total availability, externality of action, and manipulability that is associated with the traditional view of tools becomes unsustainable. Another reason why this view needs to change is the analytical perspective implied by the concept of tool, where each device is seen as an independent entity, rather than as part of a unified system. This approach was already problematic when technology was analogue, as the principle of similarity and persistence in the continuum of objects made it possible to assimilate technology to things in the world that had a single disposition of use (Dreyfus, 1992). In reality, different devices always maintained levels of interaction and mutual dependence, so it would have been more correct to read them systemically.

However, once again, we must acknowledge that the digital revolution has not revealed but rather emphasised the importance of viewing technologies as a complex system that cannot be reduced to individual tools, let alone understood analytically (Bojanova, 2014). The digital revolution has made a significant leap in the systematisation of different devices, resulting in their dialogic and interface capacity in constructing an alternative world with multiple functionalities (Sadin, 2015). The concept of the Metaverse, which has transformed from a cyberpunk fantasy to an immersive and inhabitable technological reality, exemplifies the relevance of the systemic dimension or *technospheric bubble* that renders the grounds of the concept of tool completely obsolete (Davies, 2022). While systematicity was previously hidden by the apparent disjunction or different allocation of devices, it is now apparent that it is no longer possible to separate different technologies because they are interdependent and embedded within an inextricable network. This gives rise to the concept of the *technosphere*, an ecological niche that defines the agency of human beings and the kind of relationship they establish with their surroundings. Indeed, if somatic adaptation is a correlation to one's niche, the effect of the device on the body is evident (Marchesini, 2017).

Similarly, the somatic dimension also needs to be conceptually rethought. This is not an easy task, as we are used to viewing the body as a fixed structure rather than an entity that is formed in conjunction with the external world in its becoming (Pigliucci and Muller, 2010). We are still stuck in two forms of determinism: 1) the essentialist or genocentric view, which sees the somatic dimension as predetermined by hereditary information and therefore resistant to external events, and 2) the idea of the *tabula rasa*, which conceives of the body as a completely malleable entity that takes shape from environmental factors, like a liquid that takes the form of its container. Even in a combined reading of these two determinisms, the body still appears as a passive entity that is shaped by external forces: heredity and environmental stimuli (Biuso, 2020). Instead, to understand the process of somatisation, it is necessary to adopt a dialogical conception of the body, viewing it as a proactive protagonist in its plasticity, able to produce new predicates through its relationships (Varela et al., 1992). The somatic dimension is an actively copulative entity, one that realises itself through dialogue. It can do so not due to some internal absence or insufficiency, but because of an intrinsic redundancy that establishes a degree of virtuality. Plasticity allows for phenotypic variation, which is very different from fluidity or the concept of *tabula rasa*, and should be viewed as a space of evolvability (Wallace, 2021).

Every device is received by a body that actively predisposes itself to it and together they construct a correlation. However, this does not negate the possibility that hybridisation can also have traumatic consequences on the body. Rather, *somatic phenomenology* implies that the body has an intentionality, as posited by Franz Brentano, which is expressed by relying on contents that are external but necessary to it (Textor, 2017). If the body were a text, we could say that the world constitutes a set of notes and bibliographical references that specify its content and make it intelligible. In other words, the body must be grounded in the world, as somatisation is always a reciprocal process of welcoming the device and projecting oneself into it. We must consider the implicit contents of a body not as information that rigidly and pre-deterministically establishes a morpho-functional architecture, but rather in terms of copulative hooks to the world and contents of dialogue (Gilbert, 2001). The complexity of the body makes dialogue possible, which leads to variation and, finally, unpredictable emergences. This requires a certain gradient of virtuality, a strong dose of degeneracy, a presence of iterative elements to be allocated in various ways, and a redundancy of content. Such a system presents many possibilities for expression and therefore introjects external information.

In line with the concept of polyphenism, it can be asserted that the genotype, and by extension the body, possesses a *morphopoietic virtuality* that allows it to adopt various phenotypes within a range of possibilities, based on external factors (Pigliucci, 2001). The body, therefore, is a dynamic entity that operates within threshold values – it's a plastic condition rather than a liquid one – possessing *its own characteristics* while being endowed with virtuality. Every form it assumes is an expression resulting from environmental co-factors (Waddington, 1959). The somatic condition should be regarded as a metastable entity in its morpho-functional coordinates, constantly engaging in dialogue with the external world and all elements contributing to its expression. The organism oscillates within a range of possibilities and reacts to environmental induction, by modifying its structure or modulating its functions. In this context, the device intervenes between the body and the subject's space of action, transforming both: 1) the body takes on a different configuration, bringing forth not only a new performativity but also a different perspective; 2) the space of action undergoes a joint modification, as the device constructs a proxemic bubble around the body, serving as a buffer environment that establishes its own variables through modulations and productions.

The proxemic bubble produced by a device is not an exclusively human characteristic, as all species construct a transitional space or, if we prefer, a personalised micro-environment around their body: the niche (Odling-Smee, 1988). This personalisation, or microclimatic coherence, is not properly an area of adaptive comfort, but a *space of agency* that allows the species to realise its own specific lifestyle. The adaptive view still includes the conception of evolutionary passivity of living beings that must simply adapt to their environment; in reality, the body proactively constructs its surroundings, dialoguing with the opportunities present in a certain context. When we talk about adaptive habitat, we must therefore refer to influencing variables – such as temperature, humidity, light exposure, meteorological oscillation – that are tempered by the organism's niche (Odum and Barrett, 2005). Different species inhabit particular biomes – such as a taiga, a steppe, a savannah, etc. – and have been shaped to live in certain ecological conditions. But they are not completely exposed to the rigours of their habitat: rather, they are welcomed inside a niche, capable of maintaining certain constant values – think of a termite mound, a spider web, or a dam. On the other hand, it must be reiterated that the niche is not a space of comfort but of agency, meaning a transformation of the individual's proxemic area that allows the subject to express their own style (Lewontin, 1983).

The niche is like a shell that surrounds the organism and acts as a filter against external fluctuations, but it does not eliminate the species' dependence on the inhabited environment. The niche enables species-specific agency, and niche construction involves energetic processes that stress environmental variables because they seek to change its gradients. In other words, a niche increases a species' needs at the expense of the environment, thereby amplifying its thermodynamic and ecological dependencies. The technosphere also produces a proxemic bubble for humankind, which is essentially a niche that we have constructed ourselves. Today, we know that each species does not just occupy an environmental box but actively contributes to constructing its own niche (Odling-Smee et al., 2017). However, we may mistakenly believe that the technospheric niche, by building a bubble around us, disconnects or emancipates us from our surroundings. This is a fatal error. The niche personalises space by emphasising or expressing agency in a specific way, but it does not reduce dependencies on the outside world. Instead, it amplifies them, acting negentropically (Schrodinger, 1940). In other words, the niche functions precisely like the organism, which must operate in a dissipative fashion to maintain its internal order (Prigogine, 1967).

The correlational relationship between the body and the device yields somatic changes in a process called ontogenesis or ontogeny (Gould, 1977). This is a formative itinerary that extracts a particular, unique, and in some ways unrepeatable conformation from the starting virtuality, defining the individual's morpho-functional singularity. Ontogenesis comes in three particular phases: i) the embryonic-foetal phase, where the body is still an integral part of the mother and thus is subject to all the influencing variables that come from the maternal metabolism; ii) the developmental age, which extends from birth to the subject's social maturity and establishes the coordinates of somatic development; iii) the post-developmental age, as the body still maintains a certain plasticity throughout life. The embryonic-foetal phase involves an organism *in fieri*, translating hereditary content present in the zygote, but also dependent on the parent's endocrine conditions and all the factors that act on its physiology, such as nutrition or the environment (Glezen and Alpers, 1999). The developmental age is also a period of strong dependence on external influences because the growing body is subject to all the environmental factors that can affect developmental processes. Having said that, it is evident that the body continues to respond to external variables even after that phase, for example, through degenerative or pathogenetic processes (Teperino, 2020).

Ontogeny is also marked by events of strong variability, such as the individual's specific experiences, which contribute to their singularity, and apprenticeships that follow specific developmental processes within a particular cultural or technological framework. It is this latter aspect that is most relevant when discussing the technosphere. The body develops within a predetermined space that protects it from excessive environmental fluctuations, creating a bubble that produces uniformity in phenotypic induction (Bateson and Gluckom, 2011). This means that individuals within a certain cultural-technical niche receive similar developmental influences. Although the plasticity of the body could lead to significant phenotypic differences, these divergences are contained by the uniformity of the developmental environment. Thus, both the maternal relationship and the technical-cultural niche should be considered as intergenerational factors that can influence the body's development and shape its morpho-functional characteristics, acting as a transgenerational inheritance that is neither solely genetic nor epigenetic.

The influence of the *ontogenetic niche* on phenotypic expression can be attributed to three main mechanisms: 1) modulation or selective filtering of external variables; 2) stabilisation of fluctuations in boundary conditions; and 3) production of influencing factors in phenotypic variation. With respect to the first mechanism, the niche acts as a selective filter on environmental variables, allowing certain factors to pass through while blocking others, but also dampening some and emphasising others. The second mechanism highlights the stabilising role of the niche in providing a consistent and protected environment for the developing body, reducing excessive fluctuations. Lastly, the third mechanism reveals the productive action of the niche in influencing the development of the body, which has to adapt to it (Moczek, 2015).

Therefore, we can say that when subjected to the mediating action of technology, the body reconfigures itself by modifying its phenotype. Since the technologies present in a certain historical period constitute a kind of apprenticeship that standardises the experiences of young people, it is clear that their bodies will be reshaped in correlation with them (Lerner, 1984). So let us try to summarise the variables of ontogenesis. The shape of the body is determined by several cofactors, such as: i) the genetic content, i.e., the information present in the genome of a species and an individual; ii) the remaining inherited and codifying material, in terms of DNA marking or other entities capable of acting in epigenetic translation; iii) the developmental conditions and, in the case of mammals, everything that surrounds the embryo, primarily the influence of maternal metabolism; iv) the formative processes that occur during the developmental age, especially prior to the sexual maturation of the subject; v) the events that mark the history of that particular individual; vi) the ecological niche and therefore all those aspects that make a certain living environment stable, modulate the influence of external factors, and produce their own ontogenetic pressures. Devices, in their systemic niche, ultimately represent a fundamental variable in morpho-poietic declination and teratogenesis.

4.2 Remodelling the body

The interpretive shift brought about by an interactive perspective on devices is now supported by scientific evidence showing how technology alters the functional structure of the body (Fotheringham et al., 2000). Thus, it is accurate to speak of a *technophysiology* that impacts all somatic systems. Examples of this phenomenon are abundant and extend beyond mere description, since no bodily system can be considered immune to the remodelling action induced by devices – especially those we use in our daily lives (Sparrow et al., 2011). In fact, the term "device" encompasses a vast array of objects that routinely facilitate various tasks, such as materials used for household chores, medical implants, media, food preparation tools, work tools, toys, information search tools, household appliances, transportation, communication devices, and agricultural practices, just to name a few examples. The complexity of a device can vary depending on the level of innovation or the knowledge required to understand it (Loh and Kanai, 2014). Often, users are able to operate devices without understanding their structural characteristics, which is why, for example, we can play electronic games even without any programming skills.

The body is not an autopoietic and autonomous entity, nor is it something entirely subject to external circumstances (Pujadas and Feinberg, 2012). The somatic dimension is rather a *heteronomic entity*, referring to an external normative content that is not present or included in the system. To assert somatic heteronomy is to deny the body's morpho-poietic self-sufficiency: indeed, the body, which already contains a range of possibilities in its constitution, uses virtuality as openness to the contribution of external formatting. The body is implicitly dialogical and predisposed for the assumption of externally implemented variations. The plasticity of the body, i.e., its inherent polyphenism, is a fundamental quality for adaptation to environmental fluctuations and the evolvability of the species (Minelli and Fusco, 2004). By providing for different possible morpho-functional configurations, the horizon of adaptability and colonisation of a population is broadened, correlating the organism to the opportunities and challenges present. Technology can modify the body because the body is already predisposed to be shaped from the outside.

In nature, there exists a dialectic between demands for uniformity and differentiation, which is ensured by all morphopoietic planes, starting with genetic variability ensured by sexuality, crossover, and mutations, and ending with the influences the body undergoes in the ontogenetic phase. On the other hand, it is possible to play on both fronts by keeping the morphopoietic system in a condition of virtuality, which provides for variation operated from outside (Levins, 1963). Polyphenism implies variability of individuals within the same species, not due to marked genetic differences, but to the influence of boundary variables. At the same time, this potential variability can be contained through the stabilisation of an ontogenetic niche or emphasised by making the niche a magmatic entity with a high rate of change (Laland et al., 1999). This is what is happening with the advent of new technologies. If the development of the technospheric niche in human beings was all in all quite stable until the industrial revolution, thereafter we have witnessed a trend of change at an exponential rate, leading up to today's condition of high evolution and instability. This suggests that the human body is also destined to develop different phenotypic variations in response to technological changes.

In nature, we observe polyphenisms that are related to temperature, season, nutrition, and physical gradients. Thus, the action of the niche is of fundamental importance in reducing or correlating the phenotypic variability of a population to specific contingencies (Simpson et al., 2011). If we consider the technosphere as an ontogenetic niche, its morphopoietic significance in modulating the influences of the environment and producing specific influential variables is evident, again in a logic of correlation between the organism and its surroundings. Technomediation can change the use of an organ and even the symbiosis of the organism, resulting in indirect polyphenism, as the microorganisms living within the body are strongly influenced by the niche environment (Douglas, 2014). Examples of this include the impact of food preparation techniques, personal hygiene and vaccination practices, spillover related to domestication and animal husbandry, the massive use of antibiotics, and the metropolitan environment. Therefore, regarding devices as amorphous entities that merely enhance our ability to operate in the world means misunderstanding their impact on the body, neglecting the niche action they entail. Taken together, in fact, various techniques and technologies form a developmental environment that shapes the body's morpho-functional structure. This influence cannot be regarded as a specific effect on an organ or tissue, as the development of an organism is always systemic: every variation in one apparatus modifies the embryonic development of others. For example, this is evidenced by the reduction of masseter muscles in humans due to a genetic mutation, which in turn made the development of the neurocranium possible (Carroll, 2005).

Every organ and function in the body exhibits a gradient of development and induction of activity based on the environmental stresses it receives. According to the *law of exercise*, the development of a tissue, organ, or apparatus occurs in direct proportion to the degree of activation of the related biological function. A technology can act on certain functional domains in various ways by modifying their exercise, which can transform somatic morphology and physiology. Some of these transformations have a good degree of reversibility, where the organ-function regains its organisation once the inductive factor is removed, but more often than not, this is not the case. Therefore, it is reasonable to say that all technology is embodied, and it must be considered infiltrative. For instance, our muscles reflect the daily efforts we make, and the effect of exercise is evident as the muscular structure shapes the body's conformation. Indeed, bone tissue also changes its density based on the functional stresses it receives, although its effects are not visible except later in life when we suffer from osteoporosis (Chirchir et al., 2015). Some apparatuses, such as the nervous, immune, and endocrine systems, are particularly sensitive to the action of exercise or rest that can be induced by a specific technology,

The technospheric niche, which includes pharmacokinetics, vaccinations, probiotics, and food production techniques such as pasteurisation and fermentation, can reshape the body and redefine its development. This can result in an increase or decrease in the size of certain tissues or organs, or even a more radical and extensive impact that changes the conformation of the entire organism (Lieberman et al., 2004). For example, alterations in posture not only affect certain areas of the spine, such as the cervical area, but also have effects on the pelvis and visceral organs (Cregan-Reid, 2018). The body exhibits a certain degree of plasticity, so corrective action can usually restore the normal anatomical and functional condition. However, it is important to understand the variables of the inductive mechanism to achieve stable outcomes. These variables include: i) the timing of the stress operated by the device; ii) the degree of trauma inflicted on somatic morphology, requiring compensatory mechanisms; iii) the duration of the functional stress on the organ; iv) the compatibility with the possibilities provided by polyphenism; and v) the radiation of the stress mechanism on several morpho-functional systems or its concatenation with cascading mechanisms.

When we say that an organ can restore its normal structure, we need to be very careful because somatic structure does not conform to an essential pattern, and the genome does not establish a normotype, but rather a range of possible forms that are realised differently based on external influences. This is especially true during the embryo-foetal stage, but not limited to it, as the body retains a certain degree of plasticity throughout its developmental age. Environmental factors play a crucial role as *directing agents* in shaping the morphotype in one direction or another. This ability of a given genome to generate various somatic organisations has been termed *phenotypic plasticity*, and we can regard it as the organism's susceptibility to particular *environmental releasers* during the developmental stage that induce morphological, physiological, and behavioural changes in the individual (Skinner et a. 2010). As noted above, this phenomenon has been widely demonstrated throughout the animal kingdom. For instance, the development of larvae in social hymenoptera is regulated by the food factor that determines their fate as sterile or fertile individuals (Kamakura, 2011), and the sex of tortoises is determined by the temperature at which the embryo develops inside the egg (Crain and Guillette, 1998). We should not consider these factors as accidents or deviations from the normal developmental pattern, but rather acknowledge the fact that the organism's developmental journey already includes the co-factorial intervention of the environment (Nijhout, 1999).

The involvement of the environment in the developmental process is not accidental but an integral part of the individual's morphotype – we could say that the conformation of DNA itself predicts and implies a contribution from external factors. Phenotypic plasticity, the property of giving rise to differently shaped organisms starting from the same genome according to environmental variables, is also a key topic in the treatment of many occupational or pollutant-related diseases such as neoplasms, autoimmune diseases, and infertility. Environmental pollutants have direct effects on certain characteristics of the foetus (Prutner, 2013), sometimes leading to striking consequences, such as teratogenesis, but most often inducing changes that are not immediately detectable yet still affect the body's characteristics. Remodelling is not limited to the law of exercise or the pathological/ degenerative modification of a tissue. It often involves global redefinitions of the entire body, such as the differential development of interconnected body parts with a modification of their structure and function (Bramble and Lieberman, 2004). Remodelling can also involve a process of deletion where an entire function loses adaptive value, not because technology does the same job, but because the advent of that device makes the entire functional chain no longer necessary. A different use of the organ itself can also produce the involution of some functional or structural parts, operating a remodelling, as in the case of the practognostic features related to the motor organisation of manipulation.

To summarise, let us attempt to illustrate the processes of body remodelling through specific niche dynamics involving ontogeny, taking into account the systemic constitution of the technosphere and its equally systemic effects on the somatic dimension. In short, we can categorise them as follows: 1) *rebalancing of the systems*, through differential growth, due to the exercise of some functions at the expense of others; 2) *reorganisation of the functional structure*, which occurs by disconnecting certain physiological networks and establishing others; 3) *modification of morpho-functions* depending on the type of work performed by a certain organ, which must adapt in a complementary manner to the device or develop competence of its use; 4) *changes in internal regulatory mechanisms*, which involves the surrogate action of the device and thus develops a form of dependence on it. A few examples will help understand how inadequate the concept of tool is to interpret and explain the complexity of the device's action in configuring the body. Remodelling shows us that, despite what we might believe, technology has a more profound impact on the body than on the external world.

4.2.1. The rebalancing of the systems

To understand how technology is able to shape the body during the ontogenetic process, especially during pregnancy and during the child's developmental years, we must refer to the law of exercise, otherwise referred to as the law of use and disuse. In the developmental stages, the body is not yet fully defined but in the process of growing and is therefore very sensitive to exercise stresses (Wolff, 1986). These stresses can be produced by an increase in four factors: (i) functional recall, which involves making a certain effort or being induced to express a particular behaviour; (ii) functional viability, which involves having wide expressive availability in certain areas of operation; (iii) stress or challenge, which urge the body to reinforce certain performances in order to achieve given goals; and (iv) masking or exemption, which can hide or reduce the importance of particular functions.

Technology can act on these factors by increasing or decreasing exertional demands. It is clear that technology can both increase and decrease the recall, viability, or exertion of certain functions, thereby affecting the gradient of development of a certain organ or system (Davis, 1867). Moreover, the developmental resources of a body are limited, and the metaphorical blanket is always too short, so if we pull too much on one side, we will uncover the other. We should speak, then, of an *evolutionary differential*, that is, a reallocation of growth resources. When there is a functional oversising of an organ, even if not necessarily related to a decrease in the contralateral organ, there is always a specific investment that takes resources away from other lines of growth.

On the other hand, the device often interferes with developmental processes, replacing an organ from a functional point of view and consequently mitigating its exercise (Corruccini and Beecher, 1984). The result is a consistent decrease in the organ's ability to perform a certain task. It is widely understood that the muscular apparatus grows as a result of exercise induced by a sporting activity, a job or general lifestyle; few, however, realise that it works the same way for all bodily systems. For example, the immune system is configured on the basis of the antigens it encounters: in other words, thanks to foreign microorganisms and the toxins they produce (Edelman, 1972). Antigens thus function as body building machines: the vaccine mechanism highlights this very clearly. When we speak of the ontoplastic effect of *techne* we refer to the fact that any device produces a different training of body parts. Sometimes over-exertion leads to an impairment of structural and functional integrity; sometimes, on the contrary, there can be a drop in utilisation or general underutilisation, which can be traced back to the device's functional replacement of the organ (Prakash et al., 2016). A few examples will elucidate such a deletion effect: the advent of the calculator has reduced our computational capabilities, just as the cell phone has collapsed our mnestic performance. The massive use of digital devices has caused a drastic reduction in certain cognitive functions while simultaneously increasing others (Tapscott, 2009).

Therefore, it is essential to consider the effect of technology in terms of an evolutionary differential, avoiding a one-sided focus on the positive or negative impacts of a specific technological innovation. The device's specific exercise action shows how the body must adapt to its presence through a process of complementation, creating a symbiotic relationship where one reflects the other. The somatic structure can be said to grow by filling those spaces where the organism is stressed, and this process is what I have called expression. Similarly, the process of exoneration should not be seen only as an induction of degrowth because, in many cases, it releases functional resources that can be redirected to other fields of application. Even at the phylogenetic level, we can think that a reduction of some performative skills, thanks to the surrogate action of technologies, has allowed functional reallocation and the development of other morpho-physiological dimensions of the body. For instance, we can think of the importance of techniques for cooking food or the use of tools to crush its texture (Lévi-Strauss, 1969).

4.2.2. The reorganisation of the functional structure

Remodelling can also occur by modifying the internal organisational information of the somatic system. In this case, we witness the device's ability to disconnect some functional pathways and create others. Physiology is primarily an organisation of functions that are activated synchronously or sequentially and define trends of self-regulation or autocatalysis through feedback mechanisms. For instance, hypothalamic releasing factors act on the adenohypophyseal tropins, which, in turn, act on the other endocrine glands (Molina, 2018). The hormonal products of the glands can then feedback to the hypothalamus. The endocrine life in the body system is based on cascades of events that build complex recursive chains. A technology's action can affect its normal configuration, altering its resultant arrangement and producing paradoxical effects that modify the body's metabolism (Diamanti-Kandarakis et al., 2011). Indeed, the organism constantly performs homeostasis actions to maintain optimal gradients and autocatalysis mechanisms to emphasise certain adaptive responses: these processes take place only by virtue of a functional network that must have a certain configuration to achieve its tasks.

The device inevitably produces a reorganisation of the somatic system by changing response thresholds, allostatic and homeodynamic conditions, and ushering in new performative and adaptive needs. The functioning of the body, in fact, is never given by an analytical performativity of the different organs, but by a continuous physiological dialogue between them. Therefore, we cannot simply speak in quantitative terms of balancing but we must also refer to internal functional organisation (Schulkin, 2004). A device is embodied also because it changes this functional organisation in two ways: (i) directly, by inserting itself into the meshwork of the system and bringing in new information content or functional concatenations; and (ii) indirectly, by producing new homeostatic and autocatalytic needs, which require revising sequential and recursive mechanisms. The device thus modifies the organisational information of the somatic system and affects the concerted functions that depend on the interactions and physiological sequences between the different apparatuses.

The device can transform the body's network by creating activation links between previously disjointed structures/functions, such as the arousal and gratification systems in a video game (Anderson et al., 2003). In this case, the device acts as a connective that synchronises the activation of the two functions. This phenomenon can be observed in the synaesthetic and choreographic coordination present in all arts and crafts techniques, which not only rely on devices but also require the introjection of specific sensory alliances and psychomotor scripts. In essence, a technique can be defined as the acquisition of a new configuration of the body's sensibility and agency. And let's not forget that what we see before our eyes is only the physicality of the response. We do not witness the complex events of structural reorganisation that occur in the neuronal wiring of the performance, the sensory redefinition of timeframes, and endocrine reprogramming. Therefore, any technique requires not only motor coordination, but also a metamorphosis of the functional network, involving multiple apparatuses as they are repositioned in physiological sequences or synchronicities.

The device can be considered an interface reprogrammer, acting as a mediator between the body and the world, as seen in the use of a knife, or between an organ and another tool, such as with a screwdriver (Verbeek, 2015). In such cases, the organ operates in conjunction with the tool, creating a somato-referred interface, such as the handle of an axe, and an operational interface, such as the blade. The body performs a prehensile task to achieve an operation that is not directly traceable to the implied praxis (Floridi, 2014). In these situations, the body repurposes one function to adhere to the tool's intended function. Thus, reorganisation also involves operational exaptations that create new coping resources. The association of previously disjointed functions or organs into new performative connections is a common occurrence (De Kerkhove, 1998). At other times, the device may have a dissecting effect, separating the previously synchronous functionalities of two apparatuses, potentially leading to new complementarities. For instance, video games separate the visual perspective on movement from the locomotor apparatus and relate it to the hands (Fischer et al., 2009).

The body's functional output is determined by its connectionist grammar, meaning that to achieve a specific performance, the right connections need to be established within the organism. An example of this is the construction of the brain's pathways, which occurs during the earliest prenatal experiences. In the brain's connectome, coactivation leads to the formation of neuronal assemblies that later form a single circuit of neurobiological activation (Dehaene, 2020). In the nervous system, unstimulated neurons and synapses undergo pruning during developmental stages, while their excitation leads to the growth of dendritic branches, an increase in postsynaptic receptors, a thickening of the myelin sheath, and an enrichment of presynaptic vesicles of neuromodulators. Technomediation can remodel the connectome by either increasing or containing this activation. Technomediation also deeply affects other somatic systems, such as neuromodulation pathways (e.g., noradrenergic, serotonergic), which are highly sensitive to experiential feedback. For instance, the dopamine system linked to gratification and expectation has seen a significant increase in connections following the emergence of digital media (Han et al. 2009). In any case, it is clear that every device can cause functional shifts in the organs, transforming their internal organisation, acting as a connection promoter or altering homeostasis mechanisms and sensitivity thresholds.

4.2.3. The modification of the morpho-function

Devices can also modify the way an organ works, as seen with the hand. Indeed, the hand has gone from cubital-palmitic prehensility to oculo-manual coordination in technical use, then to the flexor agility of the wrist-elbow joints in handicrafts, and later to the pincer grip from schooling, and now the synchronous functionality of the two thumbs in smartphone use. In this case, the device induces a new way of using the organ through the work involved, causing the organ to assume a new conformation consistent with the function required of it. Instead of an extension-amputation of the organ, we must speak of a *functional shift* that imparts overall conformation coordinates to the body different from those previously in force. The plasticity of the hand, which has developed in human phylogeny due to the exoneration of the locomotor task, has acted as a counterbalance to the large number of devices and techniques that have been accumulating in the cultural arsenal of our species, and even earlier in the *Homininae* (Higham, 2021). It's also worth noting that human societies have experienced a division of labour since the Neolithic revolution, leading to the production of different practognostic structures, which are now sedimented in the very conformation of the hands. Performing certain manual work implies not only technical skills but also a variation in the sensory-motor structure of the organ itself.

To understand the plasticity of the body and its organs, which we refer to as somatic expressibility, we need to revisit the concept of genotype virtuality and *phenotypic plasticity*. This means that even with the same genetic structure, different phenotypes can emerge based on two variables: (i) epigenetic intervention, such as DNA marking processes like methylation, or other components at the histone level or in the cytoplasm inherited from the parental style and (ii) ontogenetic action, brought about by environmental factors that create specific performative and experiential opportunities. A technical device alters the environment of phenotypic expression by introducing constraints and opportunities, thereby changing the morphopoietic space of the organ-function. In short, every organ presents a range of functional possibilities, and in ontogeny, the environment shapes the development of an organ conformation related to the desired performance.

The relationship between the body and the environment can be defined as a *somatising experience*, in that experiencing something leads to introjecting its growth matrices. It is important to note that ontogeny is not a rigidly predetermined process, but rather has more or less extensive margins of conformational flexibility (Weaver et al. 2007). Organic substance arranges itself by filling the spaces of experience and reflecting the context in which it is placed, similar to how the leaves of an oak tree organise a canopy that reflects the light opportunities of its environment. The use of an organ, or the particularity of the performance required of it, defines a specific developmental differential between its parts and also imposes a given form. However, ontogenetic plasticity should not be confused with fluidity or shapelessness of the organism, because there is always a limited range of possibilities to work with (West-Eberhard, 2003). Sometimes, the pathological deformation of an organ may occur when the stress goes beyond the virtual space of ontogeny. Nevertheless, the heteronomy of the body already provides for variation (Barker, 1994).

The relationship between the body and the environment is also reflected in the variability of sensory skills, which can differ greatly depending on the activities performed and the devices used. One example of this is the empowering spillover effect, where different senses are heightened differently depending on the task at hand, as seen in the differences between a sommelier, a photographer, and a perfumer (Siegel and Sapru, 2015). However, it is not always easy to recognize the shaping effect imposed by the function and metamorphosis induced by the device. Perception is a complex set of interrelated functions that can be considered a single morpho-functional structure, *shaped by the two variables of experience and performance* (Carlson, 2013). The development of digital technologies, for instance, has significantly altered people's sensory interface and the timeframe of perception.

In the context of virtual access, the sensory interface undergoes transformation not because the transducer itself changes significantly, but due to the overuse of sight and hearing and the corresponding decrease in the importance of smell and touch (Kim et al., 2012). This results in changes in the cognitive density in the brain areas corresponding to these sensors (Yuan et al., 2020). This training process shapes the interface qualities of the organism and impacts all the experiential occasions we encounter. Spending extended periods of time in front of a screen, especially during early developmental years, produces effects that reverberate throughout the organism, particularly in the conformation of the brain. In recent decades, the processing system has become increasingly static and reactive, and less trained in reflection carried out in motion, which we know underlies many cognitive processes (Carr, 2010).

Just as the natural environment affects the body's phenotype, the technosphere, through its interface function, also imposes a morpho-functional structure on our organs. For example, our liver, pancreas, intestinal microflora, subcutaneous fat positioning, blood pressure, and cardiac and vascular conformation are all defined by the type of nutrition we consume. Culinary techniques, such as cooking, salting, and fermentation, can have a significant impact on various organs, including the jaw muscles (Varrela, 1992), teeth, stomach characteristics, and bowel conformation (Campmans-Kuijepers and Dijkstra, 2021). To use another example, the use of shoes has altered the shape of our feet, sometimes leading to actual malformations that can affect our entire postural structure. Within an apparatus or organ, different types of tissues or cell populations can be modulated by a device, resulting in differential expansion.

4.2.4. Changes in internal regulatory mechanisms

When an organ undergoes morphological modification, this affects its functional arrangement and performance, which in turn can have systemic repercussions that affect other systems in different ways. A device can prompt a function or intervene in modulation by acting on the volume or sequencing of the functional cascade (Shapiro et al., 2003). While the physiology of a body has its own inherent rhythmicity, it also synchronises its clock in relation to environmental factors. For instance, the light-dark relationship regulates circadian and seasonal biorhythms in many animals, which in turn define the timing of reproduction, migration, hibernation, and diapause (Monteanu et al., 2018). Through the olfactory system and its relationship with the hypothalamus and hippocampus, real *mnestic signatures* are defined that can activate hormonal cascades, so olfactory experience cannot be considered only from the point of view of perceiving particular stimuli in the environment (Buck, 1996). Anything that affects olfactory exercise or the targeting of smell on specific molecules can have repercussions on certain aspects of hormonal rhythmicity and thus on the entire physiology of an organism.

A similar argument can be made for vision and the type of stress placed on the eyes, such as the brightness level, frequency of light, and the duration and timing of exposure (Aschoff, 1979). The retina has a relationship with the epiphysis, which regulates sleep-wake rhythm through melatonin. Thus, circadian regulation mechanisms are always in relation to the environment (Bartness et al., 1993). The device, then, acts as an interface and performs a similar function to environmental factors, by modifying the regulatory mechanisms of physiology (Drake et al., 2004). To understand the action of the device in this modulation, it is necessary to consider the concept of niche, which is the interface between the organism belonging to a species and the living environment (Flux, 1970). The niche modulates intervening variables on the body through various actions, such as: i) encrypting certain factors or providing shelter to the organism; ii) modifying values, gradients, or characteristics of environmental variables; and iii) producing other factors that intersect the body.

The technosphere can effectively be considered a niche, although it is much more complex and dynamic than those found in other species. Indeed, the niche is not merely an environmental factor or a paratypical variable, but a *hereditary element of a species* (Lewontin, 1983), similar to genetic and epigenetic factors. Indeed, just as a species inherits a genetic package of general body organisation and a set of epigenetic markings that characterise it, so it inherits a niche capable of maintaining the body within a phenotypic standard (Pechenik et al., 1998). The niche, in other words, fosters the connection between the body and the outside world, yet shields it from excessive environmental fluctuations (Mather, 1953; Lewis, 1954). The organism is predisposed to be a related entity: the principle of heteronomy indicates that the formatting contribution made by the environment is already presupposed. While it is undeniable that the organism has its own internal synchronism, such as biorhythms or functional cycles, it is also sensitive to external inductions and, most of all, requires them. On closer inspection, the body as predisposed by phylogeny already provides for *functional niche formatting* as an element of coevolution. To summarise, the body is never autonomous and always expects an external contribution (Bradshaw, 1973).

Devices can directly alter metabolism in many ways: for example, by changing the light-dark ratio, modifying nutrient characteristics, or differentiating symbiotic populations (McFall-Ngai et al. 2013). They can also do so indirectly through physiological responses like stress, fear, and arousal, which in turn activate the vagal, adrenal, and dopaminergic axis (Porges, 2011). Once again, we see the three modulation actions that are typical of a niche: encryption, modification, and production in relation to intervening variables in physiological regulation. Indeed, technology can mitigate or even replace the action of certain environmental factors, leading to a decrease or co-option of a certain function for other purposes. Alternatively, a device may change the exposure times, frequency parameters, intensity and volume of variables, or their rhythms and sequences, leading to a metamorphosis of function. Finally, there may be situations in which the technological medium itself produces inductive factors that the body must respond to, leading to the development of functions that are solicited by the device itself and must be attributed to it. In other words, we are dealing with an adaptation specifically related to the device, which is responsible for a regulatory cycle of the body.

Technological apparatuses, such as drugs, can serve as a surrogate for the organ-function, either by directly performing the same action as a neuromodulator or by indirectly inhibiting neuromodulator reuptake. The widespread use of drugs that inhibit the reuptake of certain neuromodulators, such as fluoxetine (Prozac) for serotonin or methylphenidate (Ritalin) for dopamine, can produce down-regulation with withdrawal rebound effects (Garland, 1998). It can also have a subsidiary action when an active ingredient works synergistically with a certain internal regulator or stimulates its production. Psychotropic drugs can alter physiological mechanisms, such as membrane receptivity or sensitivity of cells, gene expression, and tissue activity (Atkinson and Halfon, 2014). The functionality of the immune system can also be altered through a decreased relationship with organic material and more effective hygiene, cleansing, and disinfection. The term "up-regulation" is used when the device triggers a progressive increase in receptors, leading to increased sensitivity of cells to the action of a particular factor and physiological response (Zaliauskiene et al., 2016). Conversely, "down-regulation" refers to the opposite effect of lowering the response by decreasing the number of receptors.

Down-regulation is an interesting phenomenon because it affects the individual, who is culturally maintained within a society by sharing a particular technology. For instance, in the space of a few generations, computer media have reduced certain metacognitive abilities, such as memory, concentration, comprehension, and information search and selection due to the massive availability of data on the Web, which has become a substitute for cognitive functions (Greenfield, 2015). A decrease in demand on the organ or a lowering of stimulus load can indeed cause down-regulation, leading to a decrease in functional activation and weakening of the organ. The process of externalisation can also contribute to the decrease in functional demand. If the device replaces the organ and supersedes functional demand, we have a decrease in activation that gradually leads the organ to a slow involutional process.

4.3 Device-induced pathologies

A technology can have degenerative effects on certain organs by causing wear and tear, promoting dysmetabolic, dysplastic or neoplastic processes. It's worth noting that many pathologies have an environmental induction, and devices are among the most significant variables in pathogenesis, as demonstrated by pollution-related diseases or work-related disorders (Cregan-Reid, 2018). Furthermore, the technosphere can be considered a *microclimatic bubble* that surrounds an individual's life, directly interfering with various body functions. It is evident that many pathologies have an aetiology that is attributable to the technologies in use (Stevens et al., 2017). This is because there is no human activity that isn't aided or enabled by one or more devices. Disorders related to technical devices should often be traced back to a deficit of phylogenetic adaptation to the novelties introduced by technology: the trend of technical innovation is much faster than the mechanisms of evolutionary correlation (Travers Scott, 2018). For most of our features, our bodies are still structured according to the hunting and gathering pattern followed for thousands of years by our ancestors and other hominids who preceded them. Therefore, we might say that the problem lies in a kind of dyscrasia between technological development and adaptation.

There are numerous causal sources for the harmful effects of technology on the body, including: (i) the production of physical or chemical toxic agents that can induce neoplasms or degenerative diseases, sometimes even acting during the embryo-fetal stage and producing malformations or organic fragility; (ii) fatal or disabling accidents related to the machines used for daily activities like commuting or work; (iii) effects on the nervous system leading to conditions of stress, hypersensitivity, emotional alterations, sleep disturbances, increased irritability, and aggressiveness; (iv) severe decreases in cellular reproductive processes, resulting in impairments in the immune system and fertility, skin pathologies, and restorations in parenchymal structures; (v) severe alterations in the cardiovascular system, leading to an increase in related pathologies such as strokes, heart attacks, aneurysms, and high blood pressure; (vi) musculoskeletal alterations resulting in changes in posture, curvature of the spine (especially at the cervical level), inflammation of the intervertebral discs, inflammation in the shoulder (with bursitis and tendonitis), osteoporosis resulting from a deficit of motor activity, and arthrosis in the joints; (vii) dysfunctionalities of the gastroenteric system, including inflammation of the stomach, gastro-duodenal ulcers, persistent inflammation of the colon, and alterations of intestinal microflora resulting from food adulteration; and (viii) disorders of the respiratory system resulting from toxic or irritating fumes from motor vehicle exhaust or other forms of combustion.

The pathogenic consequences of a particular device are typically not immediately apparent, and there is often a latency period during which risks are poorly understood or underestimated (Douglas, 2020). Examples of this include radioactivity technologies, pesticides, auxins in animal husbandry, plastics, and airborne micro-particulates. We did not realise the risks until decades later, when the health problems caused by certain technologies began to emerge. The use of robotic machines in industry can also become a risk factor due to the speed and repetitiveness of the handling task required of workers, who must adapt to the pace and operational load carried out by the machine, thereby increasing the possibility of accidents or burnout. Occupational accidents are one of the main issues in the field of prevention, but unfortunately, statistics show that they are still very frequent. It is estimated that every day more than six thousand people die due to occupational accidents or diseases, causing over 2.2 million deaths a year.

However, according to the International Labor Organisation (ILO), this figure is largely underestimated. The most frequent causes of accidents in Europe and the United States involve the use of tools, industrial packaging machinery, handling of goods, falls from scaffolding, and the use of agricultural machinery.

When I talk about the stressful effects of a device, however, I am not just referring to the most obvious consequences such as pathologies or accidents, but also to alterations in the control systems of physiological stability. For example, our body has homeostatic systems that maintain certain gradients within essential parameters for different activities (Sapolsky, 2004). Allostasis is a particular physiological mechanism that achieves stability by modulating performance based on the demands placed on the organism by the environment (Bonderud, 2021). Excessive performative demands on the organism lead to allostatic overload, which has important pathophysiological consequences such as chronic hypertension and an elevation of inflammatory cytokines. This condition seems to have increased in the last two decades due to computerisation procedures, especially in the field of office work and paperwork. Today, we speak of techno-stress as characterising people who work in highly computerised environments and who often experience a deprivation of personal relationships and an excess of control over their activities (Weil and Rosen, 1997). We can define this form of discomfort through three factors: (i) stress from information overload that needs to be handled in a multitasking manner; (ii) stress from excessive use of devices involving multiple practices and applications; and (iii) stress from the performative pace and haste required in the execution of operations.

It is evident that devices impose specific performance standards on the body, requiring various levels, from the physical to the psychological, to adapt to new functional guidelines. However, the mistake we make is believing that devices only assist us in performing activities with less effort than before. While this is partly true, it is an incomplete analysis. Technological innovations change our lives, calling for a general metamorphosis that comes with its drawbacks (Brosnan, 2002). For example, automobiles allow us a degree of mobility that was unimaginable only two centuries ago, but we must not overlook the price we pay in terms of accident mortality and air pollution (World Health Organisation Report, 2018). Many pathologies that arise, especially during developmental age – such as dental alignment, altered posture, and flat feet – stem from the fact that our bodies no longer receive the physiological exercise, known as functional gymnastics, that is necessary for somatic orthogenesis. Today's children, for instance, spend too much time in front of screens, engage in little motor exercise, and assume incorrect postures, all of which has negative consequences on their development (Neupane et al., 2017).

Also, a tissue's morphology can be indirectly transformed through metabolic concatenations that favour certain chronic infections, altering immune mechanisms and the endocrine setting, and producing hormonal pathologies. The body's plasticity has precise variation ranges beyond which the system enters a pathogenic drift. Remaining within threshold values (homeostasis) allows the body to take advantage of optimal gradients to perform its activities. In this sense, there is an optimality of variables such as temperature, pH, metabolites, pressure, volume, and frequency. Homeodynamic mechanisms work like a thermostat, using negative feedback, and the threshold value, defined and monitored, determines the intervention to restore optimal variables. Pathogenic drifts, on the other hand, act through autocatalytic positive feedback mechanisms and are therefore derivative, in that they progressively worsen. When a device forcibly stresses an organ beyond its threshold values, it triggers a derivative flywheel, where physiological products themselves reinforce the pathogenic process exponentially (Ridker, 2002). In this case, metabolites resulting from stress, such as hormones or cytokines, give rise to cascades increasingly disconnected from homeostasis.

Posture is one of the most noticeable effects of today's multimedia technology, as it causes individuals to bend their necks forward and curve their backs (Portelli and Reid, 2018). This reshapes the entire body structure, leading to reduced respiratory amplitude, atrophy of the muscles in the posterior chain, and compression of the vertebrae. This is similar to the way in which certain physical jobs of the past permanently altered the biomechanics of the body. Every technopoiesis is therefore an anthropotechnique, although an unconscious one, as the relationship with the device always exerts a coercive or functional stress on the body. For example, the muscle-tendon structure of the foot, which suffered weakening due to overwork after the industrial revolution, resulting from sedentary work and standing for long hours by machines, caused deformities in other parts of the body such as the knees, pelvis, and cervical region. Today, we face a similar situation, with many hours spent sitting and looking at computer screens or cell phones. The trapezius, subclavian, and deltoid muscles are the most affected, as they become permanently contracted, leading to inflammatory conditions that affect the neck (torticollis) or the spine, especially in the cervical region (Lee et al., 2015).

Indeed, computer work is not without risks. Research shows that the endocrine system is altered by a prevalent activation of the sympathetic system, which can have considerable repercussions on stress mechanisms. Additionally, prolonged computer use can lead to a transformation of the attention system, with an increase in the reactive component and a decrease in the ability to concentrate. The wake-sleep rhythm can also be affected by overstressing the retinal-epiphyseal system, leading to alterations in melatonin production. In fact, all neuromodulation factors undergo readjustment, especially dopaminergic and serotonergic factors. It is important to note that the brain function is not limited to synaptic wiring, as reductionism may suggest. Rather, the mind emerges from the overall metabolism of the body, in which the endocrine system plays a significant role.

Consider the relationship between aggression and vasopressin, the sense of security provided by oxytocin which is produced through social behaviours, and the role of endorphins related to physical activity. These are all aspects that are being revolutionised by the new physicality introduced by digital technologies. In the case of social interaction, the role of the immune system is also crucial, through inflammatory mechanisms and the release of cytokines and histamine which can have an impact on one's mental state. We are only beginning to assess the effects that information technologies are having on people's physiology.

In conclusion

Considering the device as merely a tool used to perform given actions and only impacting the external world is a mistake. This viewpoint is limited to a single facet of technology and creates the illusion that ergonomic practices can be enough to manage somatic comfort. To be fair, this is not an entirely incorrect line of reasoning, but it is partial and therefore misleading. New technologies clearly demonstrate the dual interface function of the device, impacting both the body and the world. This phenomenon has always existed, as every technology impacts the body – now it is simply more apparent. Therefore, the interpretive model of technology needs to change, especially concerning the ontogenetic process and the relationship between the body and the environment. The concept of genotype virtuality (polyphenism) and the implicit expectation of external contribution in the developmental process (heteronomy), coupled with the new Evo-Devo and niche construction theories, show that the body is not *accidentally* modified by external factors. On the contrary: the body has constructed its adaptability by inherently contemplating hybridisation.

The technosphere is a niche that can impact not only postnatal but also prenatal development by altering the endocrine and metabolic state of the mother. Although research on the Eco-Devo effect, which examines how the niche affects embryo-foetal development, is still in its early stages, there is already significant evidence on how the growth environment affects gene translation through various epigenetic factors. Silencing or depotentiating a gene, delaying or anticipating its transcription, or emphasising its expression is enough to alter the phenotype and its metabolic impact. In other species, different environmental conditions such as the weather, or physical and chemical factors, can lead to divergent morphologies in areas such as sexuality, coat colour, adipose tissue, skull shape, and thermotolerance. In humans, predisposition to cardiovascular problems, diabetes, and various neoplasms is widely recognised. The infiltrative view of technology alters our perspective.

Sticking to the realm of morphological definition, we must not assume that the presence of a specific device, or more generally, a technologically mediated adaptive style, is limited to individual development and does not influence phylogenetic aspects as well. When we talk about evolutionary transformations, we are dealing with time spans that go far beyond historical times. However, considering the entire *Hominina* subtribe, which includes genera in the human evolutionary lineage, we see that some distinctive features point to a change in the niche (Pettitt, 2022). After separating from the *Panina* subtribe, which includes chimpanzees and bonobos, significant progressive changes occurred during the approximately 6 million-year development of this phylogenetic lineage (Higham, 2021). We can observe a reduction in the splanchnocranium in favour of a greater volume of the neurocranium and the specialisation of bipedalism, with the differentiation of the hind quarters from the forequarters. According to niche construction theory, any variation in a species' lifestyle changes selective pressures, altering individual fitness parameters within the population. We can thus hypothesise with a good margin of plausibility that increased technomediation, acting as a niche, has changed the fitness contour, favouring individuals with specific traits.

If we consider Evo-Devo theory, we can see that a mutation can have knock-on effects on morphopoiesis. A notable example is the mutation in the MYH16 gene, which produced an 80% reduction in muscle fibres in the temporalis muscle of humans (Carroll, 2005). We may wonder why this mutation became established in the human population, given that it causes a reduction in mandibular performativity, i.e., the ability to tear and chew food. My hypothesis is that the variation in fitness related to the use of fire and flint – i.e., technomediation – changed the evolutionary niche of this phylogenetic lineage, enabling individuals less gifted in terms of jaw strength to survive. After all, as Carroll points out, the anatomy of the jaw muscles has a significant influence on embryonic development, bringing us back to the theory linking evolution and development. The reduction of pressure on the mandible reduced tension on all cranial bones, allowing the skull to become thinner and larger, progressively expanding the bones of the encephalon. Thus, the technical device has had a dual impact on the *Hominina* morphogenesis: i) it changed the niche that modified fitness coordinates, allowing new mutations to assert themselves because they were more performative than the old ones, and ii) it transformed the environment of phenotypic translation, i.e., embryo development, capable of creating the preconditions for new selective competitions, which are conducted precisely on the phenotype.

Based on the examples of how devices have influenced the remodelling of the skull, the musculoskeletal organisation of bipedalism, and the transformation of oculo-manual coordination in the specialised practognosis that characterises our species, we can conclude that the use of devices has literally shaped the human body. It has not simply played a compensatory or vicarious role as classical anthropology continues to claim. We are gradually abandoning the comfortable conception of devices as obedient and inert instruments that respect the sacredness of our bodies. This is a paradigmatic shift that reflects the progressive emancipation of machines from our demiurgic role, as we shall see when discussing artificial intelligence and deep learning. We will increasingly recognize the hybridisational bearing of technology on the one hand and the emerging dimensions of partnership on the other.

CHAPTER 5

TECHNOLOGY AND IDENTITY

Premise

There is a strong connection between the traditional concept of tool and the way we perceive individual identity construction. We still hold onto the illusion that individuals have complete self-determination, and that they are the absolute protagonists of their identity formation, based on free will and without any significant conditioning factors (Steup and Ichikawa, 2018). This humanistic paradigm, which sees the individual as the demiurge of their ontogenesis, continues to inform our interpretation of identity. While acknowledging external influences, individuals are believed to have the ability to direct their lives, primarily through the power of choice (Feldman and Conee, 1985). This view is based on an assumption of ontological diversity from other animals (supposedly predefined by their nature), whereas humans are identified with the Promethean character of ranklessness and autopoietic protagonism. By considering devices as mere tools, we can render them harmless with respect to their possible co-factorship in ontopoiesis. Additionally, devices can be viewed as props that allow human beings to fully realise their own ontology, eradicated from nature and released from continuity with other living beings.

There is an implicit relationship between the universal definition of the human condition as a space of self-determination, which humanity has always interpreted freely, and the particular and singular view of an individual's ontogenetic development. Based on this presumed neutrality of the human condition, individuals are attributed the capacity to realise their own unique identities, reflecting the totality of choices and decisions they made throughout life. The making of humanity is seen, therefore, as the historical unfolding of creative processes by the individual, on a journey that is entirely internal and devoid of external debts. In this sense, any non-human entity can be converted into an instrument and used to achieve ends that the individual has conceived in autonomy. Thus, the identification of the person, in their itinerary of individuation, operates a distinction and disjunction that makes them stand out, whereby an instrumental conception underlies both the possibility of successful intent and self-determination. Identity is, therefore, placed in the frontier of *polemos*, as disjunction-distance-opposition, in accordance with the principle of individuation, which is not only free will and singularity, but also existential autarchy – the denial of all forms of co-factorality.

The dialectic between individual autonomy and the mere instrumentality of external elements leads to an essentialist conception of humanity, where its predicates are viewed as emanative rather than emergent products, given by the encounter and hybridisation with entities capable of contributing to defining the existential dimension of humankind in general and individuals in particular. This assumption is based on the idea that individuals should be considered autopoietic entities that realise themselves in total autonomy, albeit in interaction with the world. In this view, the world is never an ontopoietic *partner*, but only a tool to be used to confirm the subject's inherent predicates (Burge, 1979): there is an assumption that the individual firmly holds the helm of their existential path. This conception has traditionally characterised how the impact of technology on the somatic and expressive dimensions of humans is interpreted. Considering devices as mere materials for use that do not contaminate, but instead ensure essential purity, panders to a certain anthropological approach that is still rooted in mediaeval alchemy. It explains the yearning for purity that underlies many aberrations of the modern age. This approach downgrades the device to an artifice, trick, or expedient, rejecting it as an uncomfortable presence. What matters is the human claim to autonomy (Adams and Aizawa, 2001).

Indeed, the belief in the self-affirming role of the human being is a very useful expedient to establish a particular identity for humanity. However, this approach is limited by an epistemological assumption that requires completely inert counterparts, namely, animals and instruments. To explain the identity process that views interaction as opposition rather than hybridisation, it is necessary to assume that humans are fundamentally different from all other entities and are self-determined in their ontogenetic development without external contributions (Putnam 1975). Therefore, the anthropological machine relies on two functional operators of identity opposition: 1) animals, predetermined by nature, confined to a performative rank, and lacking free will; and 2) tools, created by humans, functional to somatic purity, but entirely passive. Both are essential to portraying humans as autopoietic and endowed with complete existential ownership. However, if the boundary between human and animal otherness turns out to be not definable, and the device exceeds its instrumental role, the anthropological edifice inevitably crumbles (Agamben, 2004).

This is just what happened in the 20th century, an era that witnessed a gradual but unstoppable process of *anthropo-decentralisation*. Darwinian theory challenged the notion of human specialness by blurring the distinction between humans and other species and emphasising the ethological basis of our behaviour (Eibl-Eibesfeldt, 1989). Meanwhile, the rise of information technologies rendered the traditional concept of tool obsolete, leading to a shift in meaning towards ontopoietic co-factorship (Clark and Chalmers, 1998). In recent decades, *scientific research on human nature* has undermined the foundations of humanism. First, the discovery of numerous extinct human species that coexisted tens of thousands of years ago has defeated the claim of human uniqueness. Secondly, evidence of hybridisation with other hominids has challenged the idea of the purity of our species. Similarly, analyses of the morphological and physiological specialisation of humans have contradicted the notion of our natural ranklessness. Finally, cognitive ethology and the increasing recognition of animal intelligence have dispelled the idea of an absolute difference between humans and other species.

A similar point can be made about the *conceptual evolution of technologies*, which are no longer seen as a crutch compensating for biological insufficiency, but rather as having decentralising characteristics, creating dependencies (Clark, 2003). In the humanist anthropological tradition, the device was seen as a support, but with the mass-media revolution of the early 20th century, its ability to widen the body's sphere of influence fully came to the fore (Dennet, 1996). As the technologies of the second half of the century highlighted their influence on the existential condition, it became clear that this impact could no longer be interpreted as an extension of intrinsic human qualities, and the hybridising significance of tools was finally understood. We recognised the ability of the device to usher in a new existential dimension, resulting from the functional convergence between the body and the technical apparatus. Finally, the so-called megashift occurred in the first decades of the 21st century, when the new AI, based on self-programming and deep learning, began to disengage from the functional design imprinted by humans, and thus from their ideation (LeCun et al. 2015).

These metamorphoses have inevitably affected the individual's space of ontopoietic freedom, revealing a less solid, autopoietic, and disjunctive identity than what was traditionally believed. Rather, human identity turned out to be hybrid and emergent, reflecting multiple constituent elements. Among these cofactors, technology has proven to play a significant role that can no longer be ignored. It is clear that identity is a construction, realised by virtue of the perceptual and operational space available to the individual as well as the social relations they engage in, in terms of the type and breadth of interactions, but also the participatory and convivial modes involved (Bechtel et al. 1998). Technology consistently intervenes in these variables, modifying the spaces of identity emergence. First of all, the person's experiential references and mechanisms change, especially in the stages of early ontogeny: I am referring to the zone of proximal growth, the secure base effect, the mechanisms of self-efficacy, and the processes of assimilation and accommodation. As mentioned, we must speak of a *technological milieu* – the technosphere – which, just like an ecological niche, defines the space of possible experiences for the individual, shaping their identity.

Identity is undoubtedly multifaceted, as evidenced by the divergence between how an individual perceives themselves and how others perceive them. These multiple dimensions of identity are formed through an interplay of differentiations and belonging (Floridi, 2014). Both technical and cultural apparatuses have always played a fundamental role in identity construction and cultural recognition. Technology produces *belonging* as an intermediary of experience, facilitating participation in a common cultural dimension characterising a given historical epoch, while also amplifying the experiential interface of the subject and thus increasing the plurality of possible outcomes. So, next to belonging, there is also *singularity*. Today, this role has become more pronounced with the advent of the digitisation of experience, which has led to increased detachment from physical relationships with things, and remote participation in social life, thereby widening the gap between different dimensions of identity. The first factor has resulted in people being increasingly enclosed within technological intermediation, while the second has led to an exponential increase in the weight of imaginary identities.

We are witnessing a real fragmentation of identity (Belamghari, 2020), with paradoxical phenomena going beyond the realm of fantasy and playing a far more substantial role than what we could have expected until a few decades ago. The invention of identity has never before had such paradoxical effects, to the point of turning it into an oxymoron. Artificially constructed online identities are developed through image apparatuses, the spectacularisation of an alleged everydayness, the accentuation of certain qualities that facilitate recognizability, and the construction of a mix of closeness, however false, and competence, albeit lacking any actual foundation. This has given rise to several phenomena, one of them being the rise of the influencer, a person who is known for being known. Using the possibilities of encryption offered by the Internet and selective self-exposure, the fictitious identity assumes such weight that it comes to affect the individual's life (Nagy and Koles, 2014). On the other hand, the Web always determines a deprivation of privacy, because while the person considers the monitor a mere window on the world through which to exercise their voyeurism, in reality, the observational potential of the Web towards *them* is far greater. The Internet produces a fluidification and transience of identity, facilitating identity theft, hacking, and ad-hoc profiling created by algorithms, with an emphasis on fictitious aspects over genuine connotations (Steven, 1997).

The relationship between technology and identity is an increasingly urgent topic to reflect upon, given the significant impact that technologies are having on the formation of the individual. Analogue devices provided experiential moments that were rooted in the reality of the world, but with the *digitisation of experiences*, the individual's living environment is undergoing a radical change. This metamorphosis of experience should not be underestimated. Digitisation is eroding the physical experience of the world, leading to a virtualisation of interaction that profoundly changes the formation of identity. It is important to carefully consider the aspects of human nature that facilitate this closure within a completely digitised world. If experience is fundamental in the construction of identity, we cannot forget that it is not only influenced by the experiential environment but also by innate propensities. Unfortunately, these propensities are often neglected due to a partial or incorrect anthropological perspective that regards human beings as devoid of phylogenetic heritage (Dupré, 2002).

5.1 Technology and the person's educational journey

The development of an individual's identity is closely linked to the available *experiential space*, which influences the psychological components through various developmental and learning processes (Rupert, 2009). This space is shaped by a combination of environmental, cultural, and social factors, which begin to influence an individual's formative process since their infancy. However, this space should not be viewed as a simple meeting point of environmental references where the individual's only interfaces are their sensory organs and agency apparatuses. In fact, *technomediation* plays a significant role in the individual's interaction with the world, acting as an ecological niche that modulates environmental interventions, defining a microenvironment of growth that tempers some variables and accentuates others. Additionally, the technological devices available in a particular historical period also modify other factors, such as social and relational ones, ultimately shaping the individual's relational interface (Noë, 2009). This phenomenon has been gradually increasing with the advent of mass media devices, globalisation, generational fashions, youth culture, and most recently, with digitisation and the rise of social media (Sweetser, 2010).

So, experience is shaped by the interaction between the individual and the world, which is mediated by the techno-cultural apparatuses that exist in a particular historical context. However, it would be a mistake to view experience as solely determined by external factors, as if the individual were a *tabula rasa* affected by environmental and technological conditionings. Instead, experience is also influenced by innate predispositions that selectively interact with the environment (Dehaene, 2020). These predispositions give rise to emotional and motivational *orientations* and establish a *hypothetical-deductive* framework. They are a part of the phylogenetic inheritance of our species and cannot be erased or undone by environmental or technological factors (Deci and Ryan, 1985). Therefore, the way an individual uses a technological device is always related to innate human characteristics. In general, identity emerges from the convergence of three interrelated factors: innate dispositions, environmental contingencies, and techno-cultural intermediation. However, these factors are not within the scope of free personal choice, so identity formation, despite being supported by technical apparatuses, cannot be fully attributed to individual agency as claimed by Prometheanism.

Both phylogenetic propensities and environmental variables, in the succession of technomediated experiential occasions, especially during the early stages of the developmental age, are not chosen by the individual but received by chance. The vision of man as the constructor of his own identity and firmly at the helm of his existential itinerary – the autonomy of those who choose for themselves – is fallacious. It leads us to superimpose man's use of devices with the Promethean paradigm of self-determination (Reza Jorjani, 2020). Ontogeny, on the contrary, is the result of *recursive dialogue between random factors*, in a historical becoming based on events that cannot be compressed into an algorithm, ultimately giving rise to a particular identity. This is not a form of genetic or environmental determinism, nor is it a simple summation of these two factors – still, the individual does not have total autonomy, but only retains partial self-ownership. Technology interposes itself as a further factor of randomness, also influenced by the variables of human nature and the occasionality of situations. Technological intermediation introduces new experiential spaces by increasing the occasions of the individual who, however, responds through the experiential endowments they possess. This means that how one interacts with technology depends not only on the particular characteristics of the device but also on the interface and epistemic characteristics that the individual has acquired from the innate heritage (Wilson and Clark, 2009).

Indeed, these predispositions define a way of exploring, selecting, interacting, and responding to environmental factors, which in turn shape our experiential orientation. Humans encounter technology through propensities such as the tendency to mimic natural phenomena, categorise and assemble objects, externalise functions through extenders and amplifiers, search internally to generate content, and imagine objects as markers of belonging, affiliation or social competition. Individuals use their a-priori representational framework, perspectival *Umwelt*, hypothetical-deductive instruments, and emotional and motivational affective dispositions to engage with the world around them and to capture the challenges and opportunities that a particular situation presents (von Uexkull, 1956). Therefore, any examination of technology must take into account the phylogenetic connotations of human experience, as it is within that dimension that our way of thinking about technology and constructing our experiential space is developed. Just as we do not approach the world neutrally and objectively, but through innate *experiential selectors*, so too technomediation cannot be solely defined by the objective performance of devices.

The technosphere inevitably influences experience because it modulates this space of agency, but to understand how experiential immersion is realised, it's essential to refer to the subjective character of its reception by humans (Kirsh, 1995). A technical apparatus fits within a system of reception that's matured by the individual, and based on this, it takes on a precise meaning: it doesn't find a *tabula rasa* or vacuum to be filled. To understand the hybridising effect, one must therefore overcome both the passive conception of the instrument that lets itself be used and the equally passive or expositional conception of the human who welcomes it: the effect is always the emergent product of the dialogue between these two terms (Clark, 1989). In fact, a device can enhance a particular human tendency or, on the contrary, mitigate it; it can make two propensities converge and produce a new expressive disposition; or it can dispense an organ from performing a certain function, by replacing it entirely. This, in turn, can cause the organ to produce a new function. In short, the effect of a device on matured identity is always systemic and cannot be evaluated as merely enlarging the operational space available to the subject.

Technology frequently exerts a far-reaching influence, especially in the early stages of its entrance onto the human stage, because it challenges intuition, bringing about a system reorganisation (Guzzetti, 2000). More often than not, a new device unhinges expectations about reality and in general the way of interpreting the external world, especially with regard to the epistemic which derives from phylogenetic endowments (Campbell, 1974). The impact is substantial precisely because it does not encounter a gap to be filled, but enters a system that necessarily has to reorganise itself to accommodate it, otherwise it would reject it.

To think of the technical apparatus in a Promethean sense, conversely, entails full human ownership of technology when, on the contrary, its impact is always reconfigurative. The telescope is not a mere extension of the eye because, as we know, its byproduct was the corroboration of the counterintuitive heliocentric hypothesis (Wolpert, 2000). If we do not take into account the collision between innate human endowments and the reorganising action of devices, we risk believing that technological development reinforces epistemological anthropocentrism, whereas it brings about the opposite effect. The identity inaugurated by technomediation is, therefore, based on a progressive decentralisation of the individual from their phylogenetic roots, a process that makes them increasingly dependent on new technical apparatuses in the desire to regain some form of equilibrium.

However, I want to clarify that my statement should not be interpreted as a negative judgement of technomediation. In fact, my intention is the opposite. By challenging the humanist view that emphasises the separation between humans and technology and reduces the latter to a mere tool, I aim to highlight the hybridising and decentralising significance of technological devices. When technology interacts with the phylogenetic epistemics of human beings, it widens the experiential space of the individual, expanding their territory and allowing them to explore previously unknown existential dimensions (Sterelny, 2003). Therefore, the effect of a device goes beyond its immediate performance, as it alters the experiential space of the subject, opening up new territories for the development of previously unimagined predicates. Ultimately, my perspective does not intend to criticise technology, but rather to recognize its potential to expand human experience and foster new forms of creativity and self-expression.

The process of hybridisation has been ongoing since the advent of the first choppers, which reconfigured human dispositions without erasing them but instead initiating a progressive decentralisation. Over time, this process produced the qualities that we now recognize as indispensable features of human identity. I want to emphasise that what we typically call *human predicates* are not the emanative result of a supposed human essence, but a co-factorial outcome fostered by the copulative propensities of human beings (Clark, 2003). They are the product of a contaminative process that has seen human beings learn from every phenomenal source, in a process that technology has reinforced. While humanism seeks the purity of the human dimension to decree its universal character, a new philosophy of technology must acknowledge that such purity is chimerical, since every human expression is hybrid (Shapiro, 2011). While humanism sees the device as an instrumental entity that allows humans to avoid all forms of contamination and somatic variation, a new philosophy of technology must recognize the hybridising effect of *techne*.

In the traditional view, human identity is believed to be independent of any factors that limit its existential autonomy, as it is assumed that (i) it has a small degree of phylogenetic determination, (ii) the device has a purely instrumental role, and (iii) human nature is disconnected from environmental variables. On the contrary, we must acknowledge that *these three factors have a significant impact on the formation of individual identity*, which, being a co-factorial result of these variables, provides a significant space of freedom. The interaction between phylogenetic and environmental factors alone ensures unpredictability of outcome, freeing the individual from any form of determinism. The intervention of technomediation has an *amplifying effect on this space of freedom* by emphasising phylogenetic predispositions and broadening the individual's experiential horizon. The subject's itinerary is unique because interaction with technology always has a destabilising effect, producing excitatory and psychagogic conditions that foster openness to new existential dimensions (Rupert, 2009). Each technology challenges the epistemic apparatus of the human being, leading to counterintuitive experiences and thereby activating an *epistemological metamorphosis*.

Bringing technomediation back to the interaction between human propensities and the experiential spaces facilitated by the device entails taking into account certain phenomena that accompany every new invention. For instance, consider the tendency to engage in an initial phase of manic and undisciplined use of the device, coupled with a low level of critical capacity, as if the new device produced a state of intoxication (Costa, 1990). The image of a person using the device coolly and rationally to achieve a previously established performance must give way to a far more complex vision. Indeed, the device acts as a disruptor in several ways. Firstly, it introduces new goals and objectives, thereby *producing* desires rather than merely responding to them. Secondly, it influences affective systems by creating genuine addictions and states of intoxication. Thirdly, it consistently modifies the epistemological structure of the human being. Finally, it influences the development of a person's identity by destabilising the horizon of expectations and transforming the subject-world interface.

In my view, placing human physiology, including its morpho-functional, psychological and expressive aspects, at the centre of the process of technomediation is a critical task that a new philosophy of technology should prioritise (Irrgang, 2021). We need to move beyond the compensatory and instrumentalist view typical of humanism, which prevents us from recognising the impact of new technologies. To achieve this, we must acknowledge the *ontopoietic competition* between somatic and technological dimensions. The body, in fact, does not merely host the technical apparatus, but is also modified by it and reacts to it, thereby influencing the outcome of technomediation. These two phenomena must be taken seriously, as they have largely been overlooked by tradition. While every ontogenetic process leads to singularities, common phenomena exist that merit attention in the study of technophysiology. These phenomena allow us to understand how the human race, with innate its tendencies and physiological organisation, on the one hand, is exposed to the influence of technology, and on the other hand, adopts the device based on well-defined predispositions (Clark, 2003).

Technology is never neutral when it comes to interacting with human beings, despite the instrumental view of *techne* that some may hold. At the same time, humans are also selective when it comes to engaging with technology. Experience is the field of interaction with the world, but it is the somatic endowment that sets the rules of the hybridising game. In other words, the perimeter defines the field, even when the outcome is decentralising (Marchesini, 2002). Decentralisation, in fact, does not involve undoing one's natural consistency, as some transhumanist literature suggests, but broadening one's horizon. The genetic inheritance defines the type of subject-world immersion, including guidelines of orientation and epistemic tools, strengths and weaknesses, propensities and idiosyncrasies, resiliences and exposures. The encounter with the technosphere produces a variation of the ontopoietic virtuality, generating new perspectives from the range of possibilities. Identity is played out on multiple levels of relationship with the world, which are established on the proxemic margin that the individual has developed (Piaget, 1970). It is a continuous process of internal organisation that is realised through external interaction. Technology functions as an intermediary and a factor of experience, integrating itself with the epistemological equipment that the subject has received from phylogeny or accrued in the course of their early experiences.

The same can be said for the emotional dimension, as technological mediation also alters the domains of affectivity, and thus the way in which experiences are somatically marked. Ultimately, we should not view predispositions as limitations, even though they clearly demarcate the field of action, but rather as the necessary condition for any form of ontopoiesis to take place. It is precisely *because* of this legacy that an individual can embark on their own journey and unfold their subjectivity. An example of this is the role of technomediation in the attachment process (Bowlby, 2015). As we know, a child does not only interact directly with a *secure base*, but also does so through objects that are exchanged and acquire *transitional value*. Soon, devices also become intermediaries in the parent-child relationship, through various affective and cognitive attunement events such as sharing experiences, focusing interests, emotional osmosis, and setting goals or desirable outcomes. Moreover, we should not forget that the universe of objects surrounding the child quickly assumes significance in fulfilling their physiological needs and channelling their motivational expressions, so devices are also introjected as integral elements of their personality (Smith, 1999). The relationship with technology ultimately influences an experience which is deeply rooted in identity and can never be expelled as if it were a foreign body, because it is part of the internal organisation of the person.

This brings me to the formative journey that shapes human identity and the question of how the technosphere, as a niche, contributes to this ontopoiesis. This development is not simply a result of exposure to the world, nor can it be reduced to a predetermined set of internal qualities that are unveiled over time. The humanist concept of autopoiesis, where identity is solely based on self-determination, proves to be fallacious once more: we are not in control of the circumstances of our birth, such as our family background, society, historical era, and cultural and technological context. However, all these factors play a crucial role in our formation and establish the foundational elements of our identity that will influence our entire existence. Technologies also contribute to our experience and identity construction by modifying the subject's experiential field in all aspects. To understand experience, we must also consider technomediation, because our relationship with the world is not direct – especially since the advent of digital technologies that immerse us in a modified reality and a revolutionised interactive praxis (Bailenson, 2018). At the same time, it's impossible to comprehend experience without considering our innate predispositions and affective ties that impact how we approach technologies. In this chapter, I will discuss three topics: 1) how innate predispositions regulate our relationship with devices, which is the other side of the technological mirror; 2) how technologies shape the way we build bonds in technomediated sentimental education; and 3) how the device defines our practices and influences the formative process, which is experience as technopraxis.

5.1.1. The other side of the technological mirror

Individual experience is influenced by multiple factors that shape one's self-image. The first aspect to consider is the animality of human beings as well as the meaning of their phylogenetic dispositions and the latter's relationship with technological innovation. Human beings are a result of an evolutionary process that established homologies or contiguities with other species as well as adaptations or peculiarities unique to their lineage. The analysis of innate predispositions in our species has to begin consistently with evolutionary epistemology, which considers the *a prioris* of human beings to be phylogenetic *a posterioris* (Lorenz, 1973). However, this approach has sometimes led to instances of biological reductionism, such as neo-Darwinism and sociobiology, which have failed to fully explain human behaviour based solely on innate and genetic components. Still, if moderated through appropriate correctives, evolutionary epistemology is an important key to understanding the dispositions and experiential orientation of our species (Bateson and Klopfer, 1973). After early pioneering works, specialised and rigorous studies have been conducted from the standpoint of scientific evidence on the most common displays of human communication, sexual and affiliative characterisations, motivational characteristics, affective and emotional conditions, and social characteristics. As a result, there is now a vast literature on the ethological traits of human beings (Moffett, 2018).

When discussing innate orientations towards technology, there are several trends we can identify that influence how innovations and devices affect people's experiences and educational paths (Spitzer, 2012). Here are a few examples: (i) The tendency to become captivated by new devices, losing any critical capacity and quickly becoming addicted to them; (ii) The propensity for laziness that leads us to delegate functions to devices and relinquish control; (iii) The strong epimeletic motivation that makes us see new devices as cubs to be cared for and educated, taking a 'parental' view that may underestimate or neglect problems while overestimating their potential; (iv) The predisposition to superstitious thinking, dividing people into believers and defeatists with respect to new technological developments; (v) The social component, including affiliative, participatory, or competitive needs, which turn devices into social intermediaries; (vi) The aesthetic tendency, which seeks the extreme sensations of the sublime or the reassuringly beautiful, activating the gratification system and causing addiction; (vii) Self-efficacy mechanisms, which reward processes that simplify and accelerate tasks, making individuals feel more capable and recognised for their abilities; (viii) The ease of acquiring habits, which removes the difficulty of having to make choices each time a task needs to be performed.

If we examine these eight points, we can see that technologies, and especially digital ones, have a significant impact on these predispositions. Consider, for example, the tendency to be captivated by technological novelties, with all the awe, wonder, and sense of magic that they arouse in us. There is no denying that human beings are particularly vulnerable in this regard, experiencing an initial hypnotic phase followed by the integration of innovation into habitual processes (Marfia and Matteucci, 2018). Only when we have established a state of habituation to the device can we engage in its critical evaluation and conscious use. However, digital technologies hinder this process by constantly introducing new novelties at an ever-increasing pace that has surpassed the time needed for habituation. As a result, we remain in a passive state of content introjection and style formatting.

The second characteristic is human laziness, or the tendency to delegate where effort is required and to shirk responsibility when it is burdensome. Laziness leads us to always choose the most convenient, fastest, and least demanding route, causing us to prefer the goal over the journey, the outcome over the action, and approximation over detail. Once again, technologies seem to cater to or even encourage this tendency, resulting in a progressive loss of human skills, including the disuse of mental calculation and the reduction of mnemonic abilities (Carr, 2010).

As for the epimeletic aspect of human beings, it refers to our strong propensity for tenderness in the face of pedomorphies, or baby-like features (Lorenz, 1943). This can be seen in our tendency to infantilize robots, digital assistants, and even the way we think about artificial intelligence development as akin to educating children. Such tendencies create an affective bond with these child-like technologies, making individuals more dedicated to caring for their needs and thus more dependent on them. The principle of care also leads to the phenomenon of the quantified self, where individuals use wearable devices to assess their physiological parameters in real-time (Swan, 2013).

Concerning point number four, human beings also have a predisposition to believe and often exhibit confirmation bias, fascination with magic, and superstition. Digital technologies and advertising algorithms tend to confirm our opinions rather than encouraging critical and open thinking. They also create filter bubbles, where users are presented with information based on their choices and echo chambers: individuals are part of groups that share similar views, which in turn tends to hinder the ability to dialogue (Douglas, 2020).

If we examine the fifth point, which highlights the social dimension of human beings, we can see how crucial it is for individuals to belong and affiliate with others, and how these tendencies play a crucial role in the need to conform to group behaviour and to feel socially recognised (Moffett, 2018). Social media, in this sense, are a trap that are difficult to escape because they offer individuals the feeling of participation and expression of their personality. The uncontrolled consumerism of digital gadgets also arises from the need to keep up with generational trends and to seek validation within the group. We should not forget that the competitive factor also plays a significant role in social motivations, which is the need to assert oneself through technology.

The sixth point emphasises the aesthetic orientation that new technologies implement, such as augmented reality experiences, speeding up of sequences with modification of timeframes, the dominance of certain senses over others, and *digiphrenia*, i.e., the need to be in multiple places at the same time (Rushkoff, 2013). Humans are particularly drawn to visually appealing images, especially those with high chromatic intensity, Gestalt coherence, iconic poignancy, stroboscopic effects, and emotional value. Virtual reality has a hypnotic effect on the individual because it constructs a landscape that emphasises these variables, creating an illusion of immersion in a world that is more fulfilling and meaningful than the natural one.

Let us now consider point seven. It is certain that humans amplify their potential through devices, for example, by overcoming various forms of intermediation and being able to obtain services directly, thus decreasing dependence on others. This affects the mechanisms of self-efficacy, whereby the individual becomes accustomed to having everything at their fingertips, and this is reflected not only in terms of convenience but also in the way they perceive themselves. Of particular importance is the ability to act remotely, to access information without leaving home, to have digital assistants always at their disposal, and to attend a class or visit any place without moving from their chair (Kluwer, 2011). Today's effortless globalised accessibility, affordability of the connection process, and the perception of ever-improving possibilities are all factors of self-efficacy that should not be underestimated.

As a final aspect, let us consider the advantages of process automation and the presence of high-viscosity containers such as Google or Facebook, serving as platforms for multiple services, creating loyalty and lending themselves well to shaping people's habits. In fact, we should not forget that humans are, by nature, habitual and inclined to prefer familiar and intuitive contexts rather than having to engage in random domains that require them to make choices. Digital technologies create *sheltering environments* where processes of childhood regression can easily take place (Cross, 2008).

5.1.2. Technomediated sentimental education

The ontogeny of an individual is not a homogeneous and linear journey but is rather characterised by sensitive periods during which one is predisposed to certain experiences and particularly receptive to certain events, as well as by formative processes that will go on to influence one's character. The attachment process that encompasses a child's early years defines the *internal operating models* through which the individual will deal with the adaptive and socio-relational situations that they will confront later in life (Mitchell, 2006). All research on secure bases and attachment types tells us that this process not only affects the definition of relational archetypes – how to deal with relationships and what to expect from others – but also influences the formation of the *autonoetic and narrative self*. This means that attachment processes influence the coherent organisation of memories into a clear self-image, as well as the experience of environmental socialisation (Ainsworth, 1978). The relationship with affective referents also includes important intermediations based on objects that become means of encounter, convivial environments, or transitional entities that sustain and renew the bond with others (Winnicott, 1971).

Thus, a child is immersed in a relationship environment and becomes accustomed to considering relationships not only as an interchange of glances and caresses but also as affiliation through things endowed with functional meaning. At the same time, the child becomes accustomed to connecting their needs with the objects around them and addressing them representatively. Initially, they experience objects orally, but within a few months they already acquire a functional conception of objects, where each is related to a specific need. The ludic dimension plays a fundamental role, creating a strong link with learning processes and cognitive resources that define a person's habitual universe in terms of search images, appraisal and coping tools, body schemas and kinaesthetic choreograms (Thelen and Smith, 1994). The direct relationship with objects and the intermediations that different devices enable produce a proxemic identity, which is a profile of the self that is defined through *connecting structures*.

Thus, we would be mistaken to read ontogenetic identity as an itinerary of disjunction from the world aimed at bringing out an autonomous and self-defined self. The formative journey of the individual, from the neonatal stages, is rather a process of introjection of the world, in which all supports that interpose themselves in the relationship play a fundamental role. Ontogenesis is an event of rootedness, similar to what happens in a plant: it is not only a matter of groundedness in the environment, but also of construction of identity-producing symbioses (Arnett, 2005). Therefore, matured identity can never be extracted by internal reconnaissance, because a being is always hybrid, constituted by the relationships that the individual has undertaken along the way.

In this sense, we speak of *proxemic identity*, which is understood as the configuration of particular relations with the world. The predicates that the subject manifests cannot be extracted in purity, as they must be referred to the relational positionality and the binding structures constituted during the ontogenetic process. Youthful experiences create a memory and *sentimental education* that profoundly affects the person's identity and self-perception (Tapscott, 1998). Proxemic identity can, therefore, be defined as a *being-with*, a way of looking onto the world through the lenses acquired in life. It is evident that the experiential milieu defines not only the territory of the known, but also *that through which new experiences can be made*. Experience creates real connections between the subject and the world, allowing for new experiences to take place. The universe of objects and devices exerts an influence on the subject because, like a vessel, it enables one to embark on a journey into the unknown. It is, therefore, a structure that connects the experienced with the experienceable, embodying the very evolutionary principle of openness to the world. A child who grew up within a 19th-century culture, devoid of automobiles, cinematographs, and electricity, experienced a dimension of intermediation that is not comparable to today's children. Ontogeny, therefore, is not limited to knowing something, but means *building bridges of experience* that connect the subject to the world.

As a child grows up, they introject the particular cultural and technospheric niche around them, constructing their own *existential connections*. This, which I have referred to as proxemic identity, not only prescribes performative, perspectival, and participatory inclusions, realised through technomediation, but also the conjugative structure, that is, the binders on which relationships are based (Rupert, 2009). Today's young people are literally immersed in digital intermediation, as they can: (i) engage in pastimes and amusements through the easy and free availability of a multitude of electronic games, which were once confined to arcades and required payment; (ii) download a large amount of movies and cartoons, learning to navigate the web as early as their second year of life; (iii) be in contact with any affective referent or friend through WhatsApp or other forms of live communication; (iv) participate in different types of social networks, especially from preadolescence onward. This inevitably changes their youth experience, which becomes increasingly static and removed from direct relationships with the world, ever more self-centred and filtered in its characterisations (Linn, 2004). It is evident, for example, that social media friendships and acquaintanceships nullify the spatial dimension and lower the sense of concreteness of relationships: it's as if presence dematerialised and lost its spatial location.

Ontogeny creates mechanisms of symbiosis, constructing bonds that hold the individual like a tree clinging to its root ball. This not only defines a sort of existential comfort zone (Douglas, 2020) but also allows for the kind of explorations that become possible for the subject. In my view, there is a worrisome neglect concerning the consequences of the digital niche on the *sentimental education* of children. It unaccustoms them to direct contact with nature, physicality in relationships with others, positionality in space-time, the importance of the material substrate of things, and the thermodynamic irreversibility of processes. It seems that young people today are losing these kinds of connections and assume a sort of functional totipotence, detached from grounding in reality. Their dopaminergic circuits of expectation and gratification increasingly become habituated to surrogate experiences, witnessing rather than experiencing life (Kringelbach and Berridge, 2009). This implies increased sensory stimulation, predominantly visual, combined with a dizzying drop in all forms of physicality. As with all forms of surrogacy, compulsive mechanisms related to unfulfilment and the maintenance of a state of constant yearning soon set in. I do not mean to say that digital technology should be interpreted as ontogenetic pollution, but rather that its pervasive use – this is the point! – has an undoubted toxicity that cannot be ignored (Koob and Le Moal, 2006).

Furthermore, when I refer to sentimental education, I am describing the connections made during developmental stages that collectively form a person's proxemic bubble of emotions and desires, similar to how the root ball is a constituent of a tree's organism alongside the roots, trunk, and crown. We can spatially represent experience as an extension of identity, although it would be more accurate to view it as a network of capillaries that extends beyond the skin, providing extracorporeal circulation vital to the subject's sentimental identity. Children growing up today, immersed in the digital realm, have a different proxemic extension compared to Western baby-boomers, who spent their childhood in an analogue world filled with numerous objects, each serving specific functions and requiring distinct techniques for use (Dickson et al. 2020). For millennials and subsequent generations, material objects have lost their significance, having been dematerialised and equated with connectivity itself. Baby-boomers, having grown up in a world teeming with functional objects, also had a very different experience from previous generations, who were still living in rural environments with plants and animals, characterised by different rhythms, smells, and close contact with the earth. Ontogeny is reflected not only in the mind, through the wiring of different neuromodulation pathways, but also in the musculoskeletal structure, with even the hands bearing praxic variations established by the ontogenetic niche (Cregan-Reid, 2018).

Proxemics is made up of memories, and memories are not mere traces of the past, heaps of sedimented information, but *sentimental styles* (Solms, 2021). Sentimental education is the result of the copulative contract a person forms during their developmental period, selecting certain references as representing the self and creating a sense of comfort and familiarity for the subject to lean into, similar to the experience of being in the mother's womb. These references are complementary elements to potential actions, which the individual introjects. The desired thus becomes the source from which the desiring person draws. However, if we interpret this experience as merely restrictive and see it as a matter of inhabiting one's memories, we create a misleading image. In fact, this anastomosing structure profoundly affects feeling, as every experience produces not only cognition of the world but also *metacognition of the self* (Cornoldi, 1997). Devices are interlocutors and interfaces that allow us not only to use them (i.e., to develop appropriate techniques and practices), but also to see the world through the focal lens they produce. Technological devices deform reality itself, and the kind of experiences we have with otherness changes along with our perspective and epistemological setup (Thagard, 1996). Devices train psychological components in a certain way and subjects them to a developmental differential, so every apparatus has a formative – not only performative – repercussion, especially when it affects a person's developmental stage.

5.1.3. Experience as technopraxis

The technosphere has always played a significant role as an experiential intermediary, even before the digital revolution. Analogue tools, while not immersive or connective, provide experiences that wouldn't have been possible without them. In hunting, a thruster resets the distance between hunter and prey; in agriculture, a plough creates a different relationship with the land. Later technologies, such as the loom, the printing press, the steam engine, electricity, the automobile, and the cinematograph, should not be considered only in the light of the performance they enable, because they modified *the human being's self-perception* (Clark and Chalmers, 1998). Technomediation is always bidirectional, widening not only the individual's access to the world, but also their exposure to it, that is, the horizon or depth of influence the world is able to exert on them. World War I is paradigmatic in this regard, as it transformed the human body into naked flesh subjected to machinic butchery, expressed by machine guns, howitzers, tanks, aeroplanes, and the toxic vapours of chemical weapons. This relationship changed the image we have of ourselves.

It is undeniable that the digital revolution has significantly advanced the development of a hyperconnected technosphere, marking a major milestone in human history (Castells, 2007). Digital technologies have fundamentally altered our experiential landscape, and they promise to do so even more dramatically in the near future. One of the most notable changes is the progressive virtualisation of the context of experience, in which the organisational information of objects – the "*res informatica*" that defines their matrix – prevails over their matter. This produces a series of significant effects, including the fact that everything that can be digitised will be, leading to a gradual but inevitable dematerialisation of our object referents (Floridi, 2014). Virtualisation will make these referents more accessible and less expensive, as in the case of movies or music files, but more importantly, it will move them from our homes to the cloud, making them less cumbersome and more practical. We are therefore witnessing a slow process of disaffection with objects as entities to be owned and handled. In this sense, our homes will change, becoming emptier and thus transforming our experience. This shift may be seen as a reversal of the 20th century, in which identity construction relied on objects that served as tangible signs of our lived experiences. Although we may still keep objects, they will likely function as decorations or reminders of the past rather than as central aspects of our living spaces.

We are witnessing a transvaluation of values, where things lose not only their affective and economic content, but also their meaning. The thing itself is increasingly perceived as a service rather than an object and may even be reproduced with different materials or mobilised in space through digitisation and 3D printing. Devices themselves will also become more intangible. Already mobile, they will eventually become introjected, entering our bodies. Digitisation results in the disappearance of things from the so-called real world, moving them into a *hyperconnected metaworld* where we will spend even more time than we do now (Veneris, 1990). Of course, there will always be things that cannot be moved to the cloud and that will increase in value, such as primary goods like food, water, and energy sources. Even these resources, though, may be revolutionised through synthetic production of nutrients and new agricultural practices in the near future.

In addition to that, the world of physical objects will certainly undergo a profound transformation through the Internet of Things (Hersent et al. 2012). Objects are already being given their own digital identity, which enables them to communicate with other devices and provide services to users. The goal is to make every object a smart device equipped with sensors and wifi connectivity, so that it can communicate data about itself and access other information that can affect its functionality. Smart objects are equipped with a contextual sensing system, which allows them to detect their location, monitor their functional regime, and establish relationships with the outside world. In this infoworld, objects are linked together in a system that situates their functionality based on positional detection and the aggregation of other data available on the Web (Pfister, 2011). Our experience with objects and the system that places them in precise functional dimensions, such as our home, will therefore change radically in the Internet of Things. Objects will no longer be seen as isolated entities, but as integrated devices. Each object will be connected to other devices and functions, more related to its situation, more modulable with respect to different regimes of functionality, more interactive with other devices in systemic complexes, more sensitive to circumstantial variables, and more intelligent in processing data.

Virtualisation also leads to the *externalisation of functions*: many aspects of ourselves that we store internally – such as memories, located in the hippocampus – will become external and therefore traceable (Ferraris, 2020). Our sexual orientations, the values we hold, the choices we make in various fields of interest, our habits, the places we visit, our physiological parameters, and even our genetic makeup will move from the private to the public domain. Furthermore, there will be no privacy safeguards that can withstand the new supercomputing algorithms. Our experience is already drowning under the tsunami of information, and we will increasingly feel inadequate in the face of the sea of possibilities that opens up before us. This perception will continue to grow, leading to a kind of *depreciation of human capabilities*, causing us to rely more on machines than ourselves (Lanier, 2013). On the other hand, artificial intelligence will only widen the gap between what the device does and how it does it, leading to a progressive loss of control. For example, thanks to deep learning, the operator has already relinquished the task of programming the algorithm, leaving them unaware of how a particular AI is able to perform a given task.

The experience that shapes an individual's identity is influenced by practice, and it is clear that devices impose their own praxis and, consequently, precise brain frames that we apply in various contexts. We can view this as the 'script effect' of technology, as each device habituates us to specific operational scripts that structure our internal patterns (Stiegler, 2016). The automation of practices is a process that, along with digitisation and virtualisation, is progressing at an exponential rate, in part because the automation of one process makes others indispensable through concatenations and synergies. This leads to a massive increase in planning, which results in a loss of spaces for occasionality, uncertainty, serendipity, and openness. The more we give away extensive portions of ownership over our existence, the more technology will influence our choices, orientations, and decisions. Once we feel that our digital assistants know more about us than what is good for us or consistent with our identity, we will be very easy to manipulate (Gawdat, 2021). As we become accustomed to a certain speed of processing, we will inevitably take value away from all those activities that seem to slow us down. I am not just talking about contemplation and imagination, but also about fundamental activities such as writing, translating, counting, remembering, ordering, organising, searching, and problem-solving. These activities are not only important from a performative standpoint but also have metacognitive relevance that cannot be replaced and is therefore indispensable for intellectual functions.

5.2 The fragmentation of identity

Discussing identity often entails navigating a complex landscape that is not easy to define. Identity is not the monolithic manifestation of an essence, but rather a process that is relational and positional, resulting in a multi-faceted structure that can contain even significant contradictions (Siegel, 1999). The attributes that an individual presents can only be partially attributed to inherent characteristics, as they are mostly defined by the relational interface in effect at a particular moment. A variation in the latter always produces a phenomenological shift. In other words, it is impossible to escape the feeling of being *One, No One and a Hundred Thousand*, as in Luigi Pirandello's novel (1925), because each time the self stands in relation to the world, different predicates emerge. The more dynamic the interactive condition of the subject, the more pronounced the identity fragmentation. The mechanisms that enable this identity pluralism are mainly transpersonal, as it is precisely in the relationship with others that the subject is compelled to assume different roles. These roles can often function as genuine forms of alienation, particularly due to the rise of mass society, which has made social negotiation even more anonymous and fluid.

Now, it is clear that the development of increasingly powerful and invasive media, such as radio and television, has contributed to the increased depersonalisation of identity, meaning the dropping of the conforming mask used by individuals to fit in their group. The individual, stripped of their personal role to integrate more easily into the new social model, has been confronted with a *virtuality of identity* that has allowed consumerist globalisation to take root (Bauman, 1998). This process has its origins in the transition from rural society, which is deeply rooted and resilient to change, to urban society, which is exposed to the influences of fashions and messages conveyed by mass media. Digital technologies have only amplified this oscillation, making it more explicit through fragmentation. To be clearer: while the analogue society allowed the individual to assume multiple forms while retaining a condition of virtuality, with the advent of the digital, the previous images of the self remain frozen in time and re-present themselves to the individual's gaze, leading to a *plurality of identities*.

We could say that the infosphere is the place where identity confronts the contradiction of mutability, or virtuality, and permanence, or plurality (Clarke, 1994). The Web creates the persistence of all self expressions, freezing every kind of clue or trace that the individual leaves, even if it is just by consulting a website, sharing a post, watching an ad, or following people on social media. Not to mention all the forms of *identity budding*, whether sought or endured, that become easily accessible in the digital dimension (Roosendaal, 2013). In the Web, identity can mutate, hide, reinvent itself, be stolen, hacked, distorted, reconstructed, profiled, or made up, but at the same time, it can be frozen in the images and statements we have shared over time, forcing us into a destabilising and coercive confrontation with our past. While the process of remembrance is always a reconstruction based on the individual's condition *in hindsight*, opening up a space for identity redefinition, the digital document comes with a timelessness that imprisons us in an eternal present.

The fragmentation of identity takes on multi-individual aspects that can be contradictory, and produces identity *exuviae*, i.e., scattered fragments of the self left on the Web reflecting past conditions. These fragments, however, are far from being empty shells, as they fully represent us and can reappear or be used as evidence of *who we are*, rather than *who we were*. Indeed, it is impossible to keep track of all that the Internet does with our images, since everything that ends up on the Web becomes a liquid public commodity flowing into any space it can find. These dispersed identities take on a life of their own and speak about us, but we are no longer able to control them. Additionally, everything that is digitised acquires a viral conformation and can only be understood through the *paradigm of replicators*, which can infect, multiply, and radiate over a multitude of tissues, acquiring mutations that make them more pathogenic or infectious (Dawkins, 1976). In other words, we risk being significantly overrun by the disseminated versions of ourselves, which could dispossess us of the self-ownership we still had in the analog world, where the hippocampus was able to organise and produce memories based on the present as opposed to past traces.

In line with the empiricist tradition, we are accustomed to considering identity as a continuity of memory and unity of consciousness, expressed as a self-narrative based on a linear historical becoming (Bruner, 1986). However, in the first two decades of the 21st century, this principle has been profoundly altered, partly due to the prevalence of immediacy, or the totalitarianism of the present, which has disrupted the narrative structure based on the consistency of historical becoming (Rushkoff, 2013). From a psychological perspective, identity is recognised through the ability to construct a continuity of experience, which is related to the neurobiology of the mnemonic act. Recollection, in fact, is not a mere trace, but a pattern that is actively reconstructed through memory. Every time we remember, we *rework the mnestic content* based on our newly assumed positions, constantly redefining our profile into a compact image consistent with the present experience (Bartlett, 1932). However, with new digital technologies, we are inundated with traces of the past that can deeply influence the way we perceive the future.

Identity fragmentation is a result not only of the proliferation of interfaces available to individuals, but also of new forms of *positional coercion* due to hyperconnection. For example, consider having to respond to emails while watching a movie, receiving messages via Facebook, or remembering to arrange a meeting with a friend on WhatsApp, all while sitting on the couch next to loved ones who are talking to us (Leonhard, 2016). The infosphere not only enters a person's here and now in an increasingly overbearing manner, but also imposes timings and encourages lifestyles that blatantly clash with the body's biorhythm. The body has specific biological cycles, such as the nictemeral or circadian cycle, the phases of the moon, the seasons, and the ages of life, which have been structured along the phylogenetic process and which our bodies expect. Each activity requires a specific physiological state that is defined by internal biorhythms and environmental stimuli, affecting the body's ability to concentrate, converse, work, be alert, react promptly, and be interested (Lacour and Belon, 2016). The problem is that for digital technology, *time is a constant*, and the time required to realise a function is based on an isochronous conception of becoming, which does not take into account the time needed for a particular activity.

This chronological misalignment not only produces a stressful condition but also profoundly alters the feeling of identity, in the Heideggerian sense of *Dasein*. Identity is not only an image that we communicate to others and to ourselves but is, above all, an experience that seeks coherence in its relationship with the outside world based on an affective and not exclusively rational feeling. Identity emerges from our *being-a-body*, from our mood, the somatopsychic spillovers of our metabolism, our intrinsic motivational propensities, and immune system intermediaries (Piazza, 2019). Our experiences depend on the metabolites of our gut symbionts, hormonal functions, neuromodulator activations, and homeodynamic events referred to physiological gradients of reference (Carlson, 2013). In other words, identity arises from emotions, drives, body states, health conditions, and metabolic efficiency, and these functions influence orientation, choices, decisions, and thoughts. Identity can fragment as a result of the multiplication of interfaces, as if we were placed inside a hall of deforming mirrors, and this is often what produces the digiphrenia entailed by digital immersion. On the other hand, fragmentation may also be related to the loss of the glue that holds together the subject's diverse experiences.

The two phenomena are often interlinked and mutually amplified by a recursive process. Indeed, the effect of identity fragmentation, caused by the multiplication of interfaces, can lead to somatic disorder that further reduces the subject's ability to *hold together the particles of the self* (Bodei, 2016). As the infosphere compels us to participate in multiple synchronous activities, causing our identity to explode into many autonomous splinters, the burden on the individual's integrative capacity increases, making it more challenging to integrate these splinters into a coherent picture.

In other words, we experience fragmentation not only because we are immersed in a world that produces *existential fissions* which persist as autonomous entities, but also because new technologies force us to live on unnatural rhythms and conditions that cause a misalignment of experience, *undermining our integrative capacities*. Therefore, both aspects of fragmentation should be considered: 1) the identity misalignment caused by somatic disorder effects related to new digital technologies, exposing the subject to greater tensions and the dismembering the self; and 2) the multiplication of interfaces that produces more divergent and co-present identity dimensions in the subject's experience, creating a multividual image of the person or a multiple identity.

5.2.1. Identity misalignment

An identity that is constantly connected is inevitably pulled in many directions, as if in a hall of warped mirrors, and this can lead to extreme stresses that are particularly damaging to more vulnerable individuals. The problem is not just exposure to bullying, which persists online long after the moment in which it occurs, but also tensions that easily erode psychological resilience (Greenfield, 2015). Depression, psychosis, and personality disorders are strongly encouraged by the absence of micro-reality, that is, the anchoring to the physical world provided by actual knowledge, a grounding environment, and physical interaction with reality. The Internet produces a condition of *existential shipwreck* that not everyone is able to withstand, as the fall of proxemic perimeters exacerbates one’s problems and vulnerabilities. It is as if the vast space that opens up around the individual becomes a mirror that amplifies, or forces constant confrontation with, the issues that afflict them (Wallace, 2014). Paradoxically, it is the most fragile individuals who are most likely to get lost in the vastness of the

Internet, those who lack the solid confidence of Carroll's Alice or the curious determination of Homer's Ulysses.

There is a tendency to overstate the relational and experiential space made available by the Web and digital technologies, without considering that many people – not just those who are explicitly vulnerable – need to operate within circumscribed affective spaces and may confuse access to the digital realm with their proxemic reality (Pariser, 2011). The lack of solid reference points and the feeling of a loss of control are the most common causes of anxiety experienced by individuals in this kind of digital shipwreck. However, unlike traditional agoraphobia, which tends to limit a person's participation in public spaces, in this case, precisely because of the apparent privacy of access, the individual compulsively seeks to reduce their sense of helplessness through interaction, ultimately getting trapped in a loop from which they cannot escape. Thus, we are facing various forms of *psychological misalignment* that particularly affect those lacking internal cohesion, exacerbating feelings of loneliness, helplessness, and increasing the risk of falling into addictive behaviours.

Once again, it is evident that technology is not merely a neutral operational tool but an entity that profoundly impacts the specific characteristics of human beings (McLuhan, 1964). Just as the mass media of the 20th century exploited human beings' affiliative motivations, leading to homogenisation phenomena with diverse and often problematic consequences, today's digital technologies address the *social aspects of our species*, which cannot be treated as a generic way of "meeting people" (Kristinsdottir et al. 2021). Human beings carry distinct social motivations such as competition, territoriality, affiliation, and collaboration, which support one another and always manifest the same dialectics (Marchesini, 2021). For example, (i) affiliation often arises as a response to competition, as cohesive bonds strengthen when a common adversary is identified; (ii) affiliation inherently generates expressions of territoriality, even in the defence of castes or professional roles; (iii) cooperation tends to increase when a scapegoat is chosen as a target for shared frustrations; (iv) various forms of assistance or cooperation require not only intentionality but also a deep emotional connection; we have strong parental inclinations, and compassion and nurturing instincts are crucial for us to establish genuine connections.

In this regard, misalignment does not diminish these social propensities; rather, it renders them more unstable and accentuates their conflictual nature. This phenomenon is particularly evident in online forums, where individuals are constantly exposed to tense situations due to the alteration of *social rituals* (Goffman, 2017). People often argue when they are driving because they have interactions at a higher speed than if they were walking and are therefore unable to perform the correct social rituals. In the same way, misalignment disrupts the established mechanisms of social interaction.

However, misalignment goes beyond mere interaction issues, as it primarily affects the individual's sense of self-recognition and solidity. Social propensities have evolved to navigate interactive mechanisms based on mutual familiarity and communal conviviality, which, although expanded compared to other anthropomorphic species, still allow for a finite number of participants. These interactions rely on a shared context of unspoken information that does not need explicit clarification because it is already established (Sperber and Wilson, 1986). In online interactions, misunderstandings are common, as each expression lacks the antecedents that define its meaning. The resulting social distress, particularly in terms of communicative frustration and a sense of social unrecognisability, arises from the expectation that interactions can be constructed without a shared foundation to support them (McNamee, 2021).

The presence of fragments of oneself on the Internet not only amplifies individuality but also increases the risk of misunderstanding, leading to a state of stress and anxiety. The indelible nature of these traces and their detachment from a core of solid identity further intensify the sense of existential turmoil, resulting in a loss of cohesion and coherence. Individuals are confronted with a liquefaction of self, where viral particles can either remain connected to the original trace or disperse throughout the web, creating a constant state of exposure and potential *loss of identity* (McQuail, 1993). This contributes to the virtuality of one's biographical configuration, generating feelings of instability. I firmly believe that this has consequences not only for the perception of identity, which naturally becomes more fluid and transient, but also for the perception of psychological continuity and bodily grounding, both of which are crucial for overall well-being. Misalignment leads to a rejection of one's somatic dimension, a state of uncertainty and dissatisfaction due to not recognising oneself in one's own body, which can manifest as eating disorders, an incessant pursuit of physical transformation, and attachment to fleeting trends (Sha and Dong, 2021).

Therefore, it is impossible to confine the problems of misalignment to the realm of psychological disorders alone, as they also have wide-ranging effects on the entire organism. Indeed, one of their primary effects is their psychosomatic impact (Nesi, 2020), as the aforementioned forms of discomfort can have detrimental effects on the immune system's ability to combat infections and neoplasms (Sapolsky, 2004). Furthermore, the cardiovascular system is also significantly affected, both due to the sedentary nature of compulsive online engagement and the stress it induces. Additionally, the pseudo-compensatory mechanisms that arise from discomfort, such as addictions and excessive consumption of junk food, also contribute to damaging bodily health. Lastly, there is the disruption of somatic experience, which diminishes the holistic nature of the body as an integrated whole.

In the 20th century, medicine had already witnessed the prevalence of a specialised approach, which was justified by advancements in clinical and surgical technologies. However, this approach was less capable of contextualising the suffering of an organ within the systemic framework of the entire organism. Now, the constant interaction with digital identities that have multiple configurations and little connection to the subject's immediate physiological state leads to a detachment from the organic planes of existence, resulting in significant consequences for people's lifestyles (Perlis et al., 2021). One of the effects of this detachment from the bodily condition, where the subject is expected to be constantly available for interaction, is undoubtedly a state of discomfort caused by a lack of synchrony (Innis, 2004).

Hyperconnectedness itself is inherently stressful as it demands individuals to constantly expose themselves to interactions, leaving little room for reflection, contemplation, and relaxation – moments that are known to be crucial for cognitive abilities and personal productivity, as recognised by neuropsychology. This continuous exposure means that at any given moment, even while eating or resting, we can be interrupted and spoken to. The solution might seem as simple as turning off our cell phones, but that would be a risky proposition: there could always be an urgent matter or we might miss out on an important opportunity. This means that we are never fully present in what we are doing, as our Self must always keep the door open for interruptions and dislocation – a phenomenon that has become more prevalent with the rise of digital platforms (Floridi, 2014). Moreover, there's an additional aspect to the demand of constant exposure: the expectation of *being available at all times* (Rushkoff, 2013). With the expansion of the infosphere, we can be contacted at any hour, emails require timely responses to prevent them from piling up, and we are monitored in our WhatsApp activity, where failing to respond promptly might be seen as impolite. Not only do we lose control over our own time, but we also struggle to allocate sufficient time to the various tasks of daily life. Time loses its inherent physiological nature – the time of the body – and becomes quantified as the time needed to fulfil a particular function, transforming into a mere chronological distance.

The individual's immersion in the ‘onlife’ world inevitably leads to a disconnection from the natural rhythms that we often mistakenly perceive as external to us – as belonging to nature. In reality, natural rhythms are intrinsic to the physiology of our bodies, governing our propensities, expectations, and availability, and guiding us in performing activities at the *right times* (Innis, 2004). We are not autonomous entities that experience life independently from natural rhythms and their interconnections with events. Rather, we are *heteronomic beings*, and our bodies always respond to something external, whether it's an environmental condition like oxygen levels or sunlight, or a temporal dimension such as the circadian rhythm or the changing seasons.

Phylogeny, the evolutionary process that shaped the morphostructure of our species, occurred through alignment with natural rhythms that existed in our ancestors. This alignment began with the rise of body organisation during the Cambrian period and continued through the physiology of fish, reptiles, and mammals, driven by the adaptive mechanism of fitness (Gould, 1989). Our planet’s orbit around the Sun, the gravitational effects of the moon, the Earth's axial tilt, as well as gravity, weather patterns, and the composition of the atmosphere, are all internalised within us and reflected by our somatic biorhythms. In many ways, our bodies are still influenced by the moon's phases that impact the lives of fish through tides, the seasonal cycles that dictate reproductive phases in reptiles' ectothermic condition, and the circadian oscillations that characterise periods of activation in mammals. These correlations developed in our ancestors and still persist in our physiological dimension, reminding us that *time is never isochronous for the body* (Eriksen, 2001).

As said, the stress of misalignment undermines the coherence of the self by promoting fragmentation. Indeed, in order to maintain these fragmented aspects of identity within a somewhat univocal orbit, it would be essential to disconnect from the onlife (Floridi, 2014) and create moments for introspection. Unfortunately, as explained earlier, this seems difficult to achieve. What is certain is that today it is humans who must conform to the time dictated by machines, not the other way around, no matter what ergonomic philosophy might think. One of the driving factors behind our compulsive connection to digital platforms is the need for control that they stimulate (Fogg, 2005). On one hand, we are curious to read the comments on our posts, check the number of likes we received, and gather as much information as possible. This creates a continuous flow that is challenging to break away from. On the other hand, the fear of being misunderstood or someone distorting an old statement to portray us unfavourably creates an incentive for controlling behaviours that support hyper-connectivity (Barbrook, 2007). It's important to remember that the intrinsic stickiness of these platforms, coupled with the controversial and conflicting use people make of them, amplifies the individual's need to constantly monitor what others are saying about them within the virtual community. This is because, in the infosphere, any attack can trigger a chain reaction, leading to events that are aptly referred to as "shitstorms."

Digitisation, furthermore, enables the dissection of an individual's document archive, allowing for the possibility of someone reporting a piece of news or an opinion in a biassed or misleading manner through ruthless editing. Consequently, individuals are compelled to remain constantly connected in order to refute, clarify, or redefine their positions. Digital identity is thus subjected to ongoing confrontations, entangled in nerve-racking negotiation processes, and held captive by those who (intentionally or unintentionally) misunderstand it. All of this is exacerbated by the pervasive nature of the online realm, whereby the infosphere tends to engulf every other dimension of reality (Postman, 1994). This gives rise to identity misalignment— the lack of self-recognition and the anguish of feeling betrayed by one's thoughts and emotions. Indeed, one increasingly prominent grievance is the sensation of being misunderstood, undervalued, or misidentified. Moreover, as the quantity of data grows, the risk of distortion becomes inevitable due to increased fragmentation. This unease can extend beyond stress or anxiety related to connectivity and lead individuals to experience depression or even contemplate suicide. Social media platforms have undoubtedly served as amplifiers of suffering, extending beyond the obvious manifestations of online bullying, primarily due to the inherent conflicts they engender (Perren and Gutzwiller-Helfenfinger 2012).

A phenomenon that has experienced an exponential surge with the advent of smartphones equipped with high-definition cameras is the selfie, to the extent that every occasion has become an opportunity to share snippets of oneself on the Internet. The selfie should be understood as a means of self-representation and interaction among individuals, a way through which the subject presents themselves to the community in a serial self-narration. These shots, capturing the person's everyday life, possess inherent identity value while serving a parallel social purpose. The attention-seeking nature implicit in selfie-mania is evident; however, it also conceals the fragmented nature of identity portrayed in these amassed profile photo collections. Selfies, as well as stickers, emojis, and animated gifs, exemplify the viral nature of modern-day narratives, in their maximal simplification and reproducibility. A selfie is meant to convey a specific meaning, which is why it is often retouched and pre-selected; it represents a small fragment of the self, a piece of identity to be inserted in the larger puzzle of everyone's documented repertoire shared on the Web. It is more than a mere self-display to others; it is a form of storytelling, a means of establishing connections, depicting elements of everyday life, intimacy, belonging, invitation, change, travel, or exceptional experiences. The selfie, therefore, serves as a recalling event that once again demonstrates how the Internet compels identity to manifest itself in a multividual manner, so as to captivate as many people as possible (Suler, 2015).

5.2.2. Dimensions of identity

Today, especially among younger generations, the Web is perceived as the primary realm for dialogue and engagement, a space where they strive to construct their best self-image and seek acceptance, recognition, and a sense of belonging. It is crucial, therefore, to acknowledge the characteristics of this medium, which has the ability to liquefy and virtualize the identity profile, fragmenting it into various positional snippets and dematerialising participation to the extent of fabricated identities and identity theft (Castells, 2009). Consequently, we witness the emergence of a *digital subjectivity* that posits relational positionality – the notion of being different in relation to the specific interface situation experienced in a given moment or condition – as the central criterion for identification. As mentioned earlier, this constant exposure exerts significant pressure on the individual's inner resources, as they become subject to the gaze of others, both being conditioned by it and receiving incentives for fiction, mimicry, and exploration of identity. The outcome is the formation of genuine dimensions of the self, which not only represent a particular profile of the individual but also define planes of interaction that often transcend the subject's deliberate choice (Aud Sissell, 2018).

When referring to *identity dimensions*, I am addressing various aspects, such as: (i) What an individual believes or assumes based on their experiences. (ii) What they aspire to be or perceive as the most suitable condition for themselves, aligning with their preconceptions, aspirations, or potential. (iii) The opinions and perceptions others have about them, including their social image and the degree of ease or discomfort it generates. (iv) Their actual identity, validated through accomplishments and institutional recognition, essentially their objective biographical background or resume. (v) Their manifestations within intimate relationships, encompassing private and unspoken aspects that are often suppressed but may surface in dreams, passions, sentimental connections, and moments of vulnerability. (vi) Their false pretences, where individuals portray themselves as something else, adopting Batesian mimicry or concealing their true identity through cryptic mimetism. (vii) External expectations and the desires of others, particularly those that offer gratification, preference, success, or simply affiliation. (viii) Their unrestrained expressions and the tendency to disregard personal and social inhibitions, embracing behaviours that would typically be avoided. (ix) What individuals are no longer, for example if they are sick or dead.

It is important to note that the presence of multiple identity dimensions in individuals cannot be solely attributed to the emergence of the digital age, as evident in literature and psychoanalysis. However, it is clear that the more veils and filters an individual has at their disposal in their phenomenology, the greater the complexity and fragmentation of these dimensions become.

On the other hand, while the dimensions of identity result from a process of fission and the manifestation of ambivalences or even contradictions within the self, they are never completely detached from one another. Instead, they are in constant and reciprocal relation, and are influenced by various external factors. In this sense, the digital realm, with its multitude of participatory platforms, inevitably contributes to increased fragmentation. For instance, consistent use of specific social media platforms like Facebook, Instagram, or TikTok can intensify public exposure and shift attention towards point iii (what others say about a subject), leading to continuous friction with self-referential dimensions of points i and ii or the institutional aspect of point iv. In such cases, the prevailing identity dimension is derived from external resonances, catering to the expectations of others (point vii) and pushing aside the imaginative and planning dimensions that individuals should gradually construct through internal dialogue and the example of respected models. While seeking validation through likes may initially enhance the social dimension, fulfilling the desire for acceptance and affiliation, it can ultimately lead to a narcissistic trap that distances individuals from reality. Consequently, a greater discrepancy arises between point i (what a person believes) and point iv (what a person is).

The digital experience amplifies fragmentation, giving rise to events that are often non-linear and unpredictable. While the anonymity afforded by the online realm can facilitate deception, identity theft, and the proliferation of fakes, it also leads to increased exposure. One phenomenon that has significantly escalated with the rise of social media is that of expressive disinhibitions, referred to in point viii (Suler, 2004). The explanation is straightforward: while face-to-face interactions are governed by inhibitory brakes, these restraints are less powerful when individuals are shielded by the virtual medium (Lapidot-Lefter and Barak, 2012). Furthermore, it is evident that direct social interactions benefit from meeting rituals and body language that help alleviate tensions and minimise conflicts. However, such mechanisms are lacking in online relationships. Consequently, there has been a dramatic increase in confrontational behaviour, polemical attitudes, and misunderstandings (Zeitzoff, 2017). It's important to note that the proliferation of connections and supposed friendships on these platforms is often superficial. Relying on flimsy foundations for trust encourages unwarranted closeness, lacking the necessary depth of genuine knowledge that would enable accurate interpretation and appropriate contextualisation of messages. This, in turn, fuels misunderstandings and fosters a contentious atmosphere that pervades these digital spaces, resulting in significant levels of stress, anxiety, and psychological distress.

It is often argued that the Internet, with its inherently document-based nature, resembles a *digital panopticon*, increasing surveillance and limiting opportunities to escape continuous tracking (McMullan, 2015). Consequently, individuals have fewer possibilities to hide or deceive, potentially making the curricular dimension (point iv) more reliable, as demonstrated on platforms like Linkedin. While this may hold some truth, it also raises serious concerns regarding privacy rights. Furthermore, the rise of the Quantified Self phenomenon, along with the likely future ease of genetic analysis, introduces a swarm of biometric parameters that could pose significant challenges in terms of individual rights. For instance, consider the use of such indicators by banks to determine mortgage eligibility, insurance companies to calculate policy premiums, or employers to select employees. When these parameters become integrated into decision-making processes driven by artificial intelligence, there will be limited room for negotiation. Undoubtedly, we are subject to much more scrutiny and assessment today than we were even just three decades ago, and this trend is likely to intensify in the future.

However, it is worth noting that the infosphere also provides spaces for anonymity, albeit in new and evolving forms. Identity cryptism (the act of hiding one's identity) and Batesian mimicry (pretending to be someone else), referred to in point vi, are significant phenomena that should not be overlooked. These behaviours give rise to the creation of fake profiles used for catfishing, which involves deceiving and manipulating others to extort money or other valuables. Common examples include sex extortion, where compromising engagements are followed by monetary demands, and romance scams aimed at extorting money through deceitful relationships (Harris, 2013). Moreover, the Internet creates a false sense of complete anonymity, as if the infosphere were an underground realm beyond the reach of prying eyes. This leads some individuals to navigate the web carelessly, accessing spaces and services they would prefer to keep hidden. Instead, it is crucial to recognize that nothing is more public than one's online activities. The intimate dimensions of identity, encompassing personal and private aspects (point v), typically pertain to matters confined within the home and close relationships, such as between partners. When an individual accesses a particular website from their desk, they are no longer shielded by the walls of their room but instead find themselves exposed in the virtual public square, under the gaze of a vast audience.

Indeed, every action taken within the digital realm is not only tracked but also subject to potential spying, documentation, and even playback within the community. In some cases, the situation can be even worse if one's system becomes infected with remote access malware. Remote Administration Tools (RATs), such as the infamous Trojan, named after the Trojan Horse of ancient mythology, allow external users to maintain remote control of a computer. This not only grants them access to our files but also enables them to monitor our activities, intercept our phone conversations, and potentially activate our computer's webcam for surveillance purposes. Camfecting, the act of hacking into a webcam without the owner's knowledge, is a method used for blackmailing individuals.

While there are defence measures available, this is an ongoing battle between protective systems and hacking resources. This highlights the misconception of presuming online anonymity solely because one is operating from the privacy of one's bedroom. Additionally, there are other forms of identity harassment on the internet, even within the realm of point v. One is the tendency to steal photos of children, shared by relatives, to post them in pornographic sites – this is known as digital kidnapping. Another is digital identity theft, which can be exploited for various purposes, such as creating follower packs to manipulate polls or preferences. Another prevalent phenomenon, particularly on Instagram, is the hacking of accounts to steal sensitive information, where requests for loans or favours are made using the compromised profile. Digital cryptism, often employed through the use of anonymous nicknames, serves individuals who wish to send provocative messages, incite online quarrels, or act as trolls on large-scale platforms (Wachs and Write, 2018).

In short, the Internet contributes to the proliferation of identity dimensions that often leave individuals feeling fragmented and lacking a cohesive and stable identity. As mentioned, this fragmentation has significant implications for a person's psychological well-being, particularly because many individuals are not yet fully accustomed to or familiar with the opportunities and challenges presented by these platforms. Experiencing digital bullying or being targeted by coordinated trolling campaigns in public forums can have profound psychological effects. In general, navigating the vast sea of the Internet has become increasingly challenging for individuals, as they risk becoming lost amidst its expanse, confused by the multitude of reflections it presents, and overwhelmed by its unpredictable waves (Gackenbach, 2007).

Consequently, if a platform were to thoroughly profile us based on our online activities, they would discover an abundance of documentary evidence that surpasses an individual's capacity to construct a self-profile. By piecing together data, it becomes effortless to predict future behaviours, inclinations, quirks, and even the most intimate aspects that characterise a person. Moreover, the digital realm proves to be a challenging place to hide not only from others but also from oneself. It blurs the boundaries between private vices and public virtues that we typically keep separate. As identity emerges from the interplay of identity dimensions and one's relationship with the world, any changes to the subject's interface will profoundly impact the construction of their identity. Some dimensions contribute to a more fluid identity, while others stabilise it and restrict its ability to change, such as the persistence of outdated self-images that no longer align with the individual's present state (point ix). This affects the individual's development, as they face the contradiction between a remembered past, reformulated through the mnemonic process, and the online document. As mentioned, the mechanism of memory presents recollections as reformulations, unlike digital documentation that tends to crystallise experiences, anchoring the subject to their past and reducing the potential for forgetting or reinterpreting it.

The transformation of the self into scattered replicators leads to a fragmentation of identity. This fragmentation not only relies on the situational and circumstantial nature of individual data but also follows its own evolution, resembling a virus that reproduces itself through mutation and hybridisation, adapting to different contexts. It is no longer us who shape our memories and give them a representation that best reflects our identity, but rather the opposite (Floridi, 2014). The proliferation of residual yet active digital presences on the Internet prevents us from maintaining a cohesive and coherent self-image, constantly challenging our self-view. Indeed, users cannot help but rely on what they find online, mistakenly believing they are directly interacting with other people. The misconception of direct contact between the individual and their digital avatar poses one of the numerous challenges faced by messaging platforms, email services, or apps like WhatsApp, where online presence is often mistaken for the actual availability and presence of the person.

A similar situation arises with post-mortem digital presence, namely the phenomenon of a person's digital identity surviving after their biological death, including potential posthumous online interactions (Sisto, 2022). The persistence of digital identities on the Web creates a form of incomplete mortality, a disconnection between the existence of the individual and the occurrence of their death. To comprehend this transformation of identity, it is essential to consider that an individual generates an average of a thousand online documents per year, which continue to circulate through shared posts, unread messages, and notifications of memories. The generations born in the new millennium produces a vast archive of data, far exceeding their mnemonic capacities. The recent discipline of thanatechnology (Sofka et al., 2012) is attracting numerous researchers who explore various studies and generate a multitude of theories related to the management of grief, the preservation of information about the deceased, and the potential for posthumous interaction (Ozdemir et al., 2021). It is now commonplace, for example, for a long-forgotten message posted years ago to resurface and generate new discussions and shares within online forums. The same phenomenon occurs with deceased individuals who suddenly become the focus of collective interest, inundated with inquiries and shares. What impact does this continuous revival of mourning have on friends and relatives, particularly when the memory of the person is not respected or becomes subject to distortions and unwarranted attributions?

We often speak of a new form of immortality, but from my perspective, it would be more accurate to describe it as a postponement of the sense of uncertainty associated with mortality. We cannot overlook the anxiety-inducing and stressful impact of losing control over one's identity. It is one thing to believe in true immortality, where one retains the right to respond and redefine one’s positions, but it is quite another to feel completely vulnerable to media manipulation. Presently, we can already observe a significant online presence of deceased individuals who remind us of their birthdays, share stories, and leave unread messages that seek attention. The emergence of social media has encouraged people to share their daily lives through phrases, photos, and short videos, expressing their opinions, feelings, preferences, and desires. Consequently, this vast archive of data could potentially be organised into one or more post-mortem profiles, either spontaneously or under the guidance of someone with an interest in the deceased. Therefore, we should speak of a digital spectre, comparable to a form of artificial intelligence that transcends a mere diary or fossil record, but rather represents an evolving entity with its own online existence. For example, it could continue publishing Facebook posts or responding to user inquiries. This indeed would be a new form of immortality, albeit algorithmic. And there is no reason to ignore the possibility that such an avatar could eventually emancipate itself from being a residual ghost and assume a distinct identity, potentially diverging significantly from the original.

CHAPTER 6

TECHNO-ADDICTION

Premise

The concept that the body is impermeable to the influence of technology, or at least that its impact is limited to individual organs, was favoured by the humanist paradigm (Heidegger, 1959). By viewing humans as closed-off entities with a fixed essence and technology as a mere tool, the classical understanding of *techne* hindered a comprehensive understanding of its twofold interference. Adopting a bi-directional perspective involves recognising two crucial points: 1) the device, with its characteristics of impact and mediation, actually modifies the body by influencing various somato-plastic factors at different stages of both phylogenetic and ontogenetic predisposition; 2) the world, when accessed through a device, intensifies its influence on an individual's constitution, as the technological apparatus does not screen interference but rather modifies its nature and generally amplifies it. If the humanist model of the device resembled a glove, representing a tool that separates and protects the body from the world, today's prevailing model is more akin to a chip, which *enhances* the subject's interaction with, and exposure to, external reality (Lanier, 2011).

There are also other factors that have contributed to fostering a dismissive attitude towards the impact of technology on the body. These factors include: i) the latency period that typically occurs between the use of a technical device and the recognition of its effects on the body and ii) the tendency to forget what life was like before the introduction of a given device and the simultaneous development of dependence on that device. This leads to a lack of awareness regarding the modifying effects, which can sometimes be pathogenic, brought about by different devices. This is primarily due to the habituation to performance gains: in a given historical period, people become accustomed to the benefits provided by devices, while overlooking their potential drawbacks (Eyal, 2014). It is important to note that technology, or more accurately, the *technospheric dimension* encompassing a range of technical apparatuses, is sustained by a particular societal structure that relies on the production and system dynamics facilitated by these technical apparatuses.

19th-century European society, shaped by the Industrial Revolution and the emergence of factories as workplaces, brought forth a range of problems that can be attributed not to individual machines, but to the specific *living conditions* created by this technospheric dimension. During that time, the concentration of large populations in new urban centres, characterised by inadequate sanitation, overcrowding, and high levels of air and water pollution, facilitated the spread of serious diseases like tuberculosis, typhus, and cholera (Cregan-Reid, 2018). The prolonged hours spent standing without movement resulted in severe spinal column alterations, particularly affecting intervertebral discs and joint cartilage. The feet, which rely on movement for proper function, also suffered as prolonged standing led to hypoxia and skeletal structure impairment, causing bone and muscle density atrophy (Chirchir et al., 2015). Weakened feet, in turn, disrupted the entire biomechanical posture, leading to pathological changes that affected other parts of the body (Nutt, 1915). Additionally, the regular use of heavy machinery often resulted in accidents involving fingers, limbs, or other forms of mutilation. It was not uncommon for workers to be crushed between gears – something that unfortunately still occurs today. Therefore, each technoscientific revolution gives rise to a particular type of society centred around a distinct technospheric dimension, which in turn is responsible for somatic alterations.

Paleontological evidence reveals that even the Neolithic revolution had significant consequences on the human body. It resulted in degenerative changes, primarily affecting bones and teeth, and introduced new epidemiological challenges with the emergence of zoonotic diseases caused by animal husbandry (Lieberman et al., 2004). An illustrative example of the impact of technological practices on human health can be found in ancient Roman *saturnism*. Lead, a widely used material due to its ease of processing, was employed in the production of pipes and utensils such as pots, pans, and wineskins. Chronic exposure to lead, as we now know, leads to a range of disorders affecting various systems, including the nervous system. Symptoms of lead poisoning include increased aggression, mental retardation, tremors, irritability, cerebral edema from arterial spasm, and peripheral nerve paralysis (Needleman, 1999). A similar condition called "phossy jaw" was prevalent in the latter half of the 19th century among workers in the match production industry, where white phosphorus was used. Phossy jaw refers to necrosis of the jaw bone, accompanied by inflammation of the alveolar-gingival area and subsequent ulcerations that affect the periosteum and bone. Even in modern times, occupational diseases remain a significant concern, closely tied to the technologies employed in various workplaces. Examples include noise-induced deafness, contact dermatitis, pulmonary diseases caused by dust inhalation, and chemical intoxication (Kim and Kang, 2013).

Today the main concerns regard the pervasive exposure to radiation from various devices such as mobile phones, computers, Wi-Fi systems, power lines, and microwave ovens. Since the 1980s, there have been claims of electrosensitivity, the development of tumours, and alterations to the emotional system, including anxiety and mood swings (Rubin et al., 2011). While the precise pathogenic effects of these exposures are still being studied, it is undeniable that new technologies have an impact on our physiological systems. One of the notable consequences of technological advancements, as mentioned, is the acceleration of the pace of life and the heightened performance expectations in our daily activities. This inevitably leads to a state of prolonged stress, commonly known precisely as *technostress*. However, this is not the only discomfort we experience (Colvile, 2016). The constant pressure to meet high performance standards creates a persistent anxiety about not being able to cope, being seen as incapable, being socially isolated, or failing to fulfil bureaucratic obligations. Each technological evolution establishes new performance norms and redefines the parameters of functional acceptability and normality (Dossey, 2014). One prevalent source of distress is the feeling of frustration, stemming from the inability to fully control the processes we encounter on a daily basis (Przybylski et al., 2013).

Another source of frustration in our time stems from the rapid pace of technological innovation, which often leads us to adopt devices or practices without fully comprehending their intricacies. This causes the feeling of being somewhat coerced by technology, with a corresponding sense of loss of autonomy (Toffler, 1970). The growing disparity between the services provided by today's technical devices and the way in which they operate, a gap that is widening with the emergence of new self-programmed artificial intelligences, engenders a state of passive dependence and almost fatalism towards a system that we acknowledge we can no longer control. Undoubtedly, this contributes to feelings of anxiety and frustration as well.

Another issue is the excessive stimulation caused by new digital technologies, which affects us on two fronts. Firstly, they permeate every aspect of our lives, as constant connectivity prevents us from disconnecting from work or the interference of others. This results in a loss of personal and private space, leaving us vulnerable to constant interruptions (Nippert-Eng, 1996). Secondly, the very nature of these technological interferences leads to *heightened arousal*. The brightness of screens, the sensory alerts we receive, and the rapid pace of expected responses all contribute to increased stimulation. Stress from onlife is one of the major challenges of our time. It encroaches on our privacy, granting little respite from the outside world, and fosters a compulsive need to constantly engage and interact with digital technologies.

The term "techno-addiction" is used to describe the growing forms of dependence on devices that increasingly impair people's ability to use these technologies in moderation, potentially ensnaring them in their grip (Block, 2008). It is clear that Big Tech companies, such as Facebook, Google, and Amazon, have a vested interest in keeping users connected for extended periods of time, effectively fostering genuine addiction. They achieve this by enhancing the stickiness of their platforms, offering more services and expanding the range of contexts in which they can be utilised, thus encouraging users to remain engaged (Eyal, 2014).

Another approach is to leverage the mechanisms of the *affective seeking system* by: i) manipulating motivational engagement through specific recall or proactive impulse actions; ii) creating opportunities for expressive exercise, prompting highly specific actions; iii) reinforcing expectations, gratification, surprise, or variable reinforcement; or iv) fostering a sense of investment that extends beyond the immediate activity, forging a connection between the user and the platform, thus encouraging return visits (Deci and Flaste, 1996). We are aware that the affective seeking system operates by activating certain neuromodulators responsible for interest and pleasure, such as dopamine, serotonin, and endorphins. It may seem paradoxical, but when we share a photo on Instagram, we are essentially engaging with that ancestral seeking and reward system that enabled our ancestors to forage for resources (Lichtenberg, 1989).

Another method of fostering addiction is by creating an immersive bubble for the user – a confined space of content that aligns with their interests, values, and desires, populated by like-minded individuals (Douglas, 2020). In essence, these platforms cater to users by providing an environment that reinforces their preferences, never questions them, and encourages repetitive and passive behaviour, ultimately stifling critical thinking. They often exploit the innate inclination towards confirmation bias, rather than challenging and overcoming it. Additionally, they offer users more than they express or desire, making them feel moderate by comparison and thus further reinforcing their biases (Davidson and Farquhar, 2020). Consequently, this mechanism promotes infantilism and progressively isolates individuals within their own impulses, akin to dog training techniques that employ reward-based conditioning. Unfortunately, this establishes a self-reinforcing cycle or feedback loop that deepens the person's dependence on these platforms. It's worth noting that the mechanisms of addiction are largely the same, whether we're discussing gambling, video games, or social media (Kardaras, 2016). The more a behaviour is expressed, the more it develops, increasing the likelihood of future repetition due to *synaptic reinforcement*. Furthermore, over time, the reward mechanism leads to *habituation*, prompting individuals to spend more time on their devices in pursuit of the same pleasurable effects (Berridge and Kringelbach, 2015).

On the other hand, dependence on technical devices is also linked to the phenomenon of internal regulatory mechanism modification. One well-known example of this is *down-regulation*, wherein if a device replaces a function or maintains its effects at a high level (similar to the action of neuromodulator reuptake inhibitors), the body will tend to down-regulate the physiological mechanism associated with that specific function (Nestler, 2014). As a result, individuals become increasingly dependent on the device's supplementary intervention and may perceive their own somatic dimension as deficient. This is further compounded by the fact that externalised functionality not only replaces an organ but also performs better. There are two forms of down-regulation: ontogenetic, which influences the differential exercise mechanisms of organs, and phylogenetic, which acts as niche construction by encoding certain selective factors and consequently diminishing the organ's performance within the species. The second regulatory mechanism is *up-regulation*, which involves the functional elevation of a physiological mechanism due to the stress or new functionalities introduced by a device (Sherwood et al., 2012). This regulatory process can also occur at the ontogenetic level, increasing the demands on a specific organ, or at the phylogenetic level, heightening selective pressures on a certain function. Up-regulation is a common occurrence in allostatic responses, but it can also result from increased demands on somatic performance or more intense exercise using a particular apparatus. For example, electronic games can enhance a player's selective attention and responsiveness (Fernandez-Aranda et al., 2012).

Dependence on technology is not a recent phenomenon and is not solely confined to our era, but it occurs whenever a species utilises a device. However, it is undeniable that the rapid pace of technological advancement in the last century has brought this phenomenon to the forefront. In essence, there is no substantial difference between adaptation to technology and adaptation to the natural environment; both involve a correlation between the body and the external factors it interacts with. The evolution towards heteronomy, which refers to shaping the body in response to external formatting or supports, is a common characteristic across all living forms (Odling-Smee et al., 2017). The development of new functionalities can often be seen as the repurposing of an existing function for a different task (this is known as exaptation). This repurposing is made possible either by changes in environmental conditions that no longer require the previous function, or through externalisation, which entails another entity, such as a symbiont or a device, fulfilling that particular function. Furthermore, it can be argued that the very concept of adaptation implies that every evolutionary process is, upon closer examination, a co-evolution – a symbiotic correlation between organisms and their environment or technology (Thomas, 1974).

There are numerous animal species that have been shaped by their use of various tools, but it would be incorrect to assume that species not utilising tools are exempt from somatic shaping processes. Even the relationship with the environment creates dependencies that lead to the development of techniques necessary for sustenance. The subtribe of *Hominina*, starting from the Australopithecines and especially after the emergence of the *Homo* genus, has always been characterised by the extensive use of diverse tools, with performance being consistently achieved through functional externalisation (Higham, 2021). Technological dependence has been intensifying at an exponential rate, so it is no surprise that the past few decades have witnessed the most significant acceleration of this process. Earlier I described a future in which technology increasingly permeates human existence, but it is important to recognize that the origin of this dependence stretches back into antiquity and cannot even be solely attributed to our species. For instance, *Homo habilis* already demonstrated mastery in the use of chipped tools. Therefore, we must acknowledge that our dependency on technology has always been inherent to our existence.

6.1 Have we always been cyborgs?

The concept of "cyborg", embodying the infiltration of technology within the human form and manifesting hybrid functions and capabilities that can be attributed to both the organism and technological components, gained popularity in the 1980s through cyberpunk fiction. During that era, brimming with science fiction influences, the cyborg remained either a fascinating notion or a source of nightmares, depending on one's perspective. However, just a few decades later, it began to assume increasingly common, almost mundane, manifestations in its practical applications. The integration of subcutaneous chips, electronic sensory devices, and progressively intelligent prosthetics in the early 21st century unequivocally signalled the advent of the cyborg age. If we consider hybridisation as the outcome of a *permanent graft*, akin to a form of somatic parasitism (for example, an endosymbiotic chip), we can claim that that was a true watershed moment in technomediation (Margulis, 1970). Alternatively, if we view hybridisation as a functional presence that need not be confined solely within the body, it becomes evident that the co-evolutionary metamorphosis of the somato-technological relationship had been unfolding for quite some time. This raises the question: When did we truly become cyborgs?

Each technological innovation transcends its confines and manifests itself in various forms (Barnstone, 2000). In fact, it: (i) generates disparities in exercise across different apparatuses, much like a bodybuilding machine; (ii) disassembles specific functions and reassembles them into alternative performances, leading to a sort of physiological dissection, as seen in the sensory experience of movement in electronic games; (iii) alters the functional dynamics of organs, as is the case with the transformed usage of hands with the advent of digital interfaces; (iv) induces down-regulation by externalising tasks, as exemplified by membrane receptors or neuromodulator reuptake inhibitors; (v) prompts up-regulation in some apparatuses, compelling them to heighten their functional responses, as observed in mechanisms like physiological solicitation, hormonal sensitivity, or allostasis; (vi) fosters addiction, akin to certain psychotropic drugs.

Techno-addiction primarily encompasses the ontogenetic dimension through developmental and learning processes governed by: 1. the principle of differential exercise, or Davis's law, encompassing the use and disuse of organs; 2. the law of effect or Thorndike's law, where operants are reinforced or punished; 3. the law of composition or Hebb's law, which posits that coactivating specific parts of the neurobiological system leads to their interconnectedness; 4. the law of mimesis or Bandura's law, involving the influence of social mirroring through collective activities mediated by the presence of devices.

Every factor that influences the developmental processes during the critical stages of growth and maturation has a profound impact on the morpho-functional organisation of the body. Therefore, it cannot be merely regarded as the acquisition of knowledge and skills. Hence, we can assert that hybridisation is an integral part of individual ontogeny because, right from an early age, the developmental processes that shape our physiological configuration are mediated by technology (Kardaras, 2016). Moreover, according to the latest evolutionary theories, we must also consider the phylogenetic effects, which exert influence on the morpho-poietic trajectory of the species as a whole (Pigliucci and Muller, 2010). According to Evo-Devo theory, in fact, any influence on embryogenetic regulatory mechanisms has systemic effects on the phenotype, giving rise to new patterns regulated by natural selection. Similarly, the epigenetic theory suggests that certain modifications acquired by parents alter the boundary conditions of the zygote's genome, such as histones, thereby altering the phenotypic expression. Additionally, according to niche construction theory, any transformative changes made by a species to its living environment, actively shaping selective pressures, have a morpho-poietic effect (Odling-Smee, 1988). It is evident, therefore, that these effects extend beyond individual ontogeny, entailing the co-evolution between the human body and technology. Consequently, we must recognize the hereditary remodelling of the somatic structure of *Homo sapiens* that occurred throughout phylogeny.

As noted, the set of devices creates a technospheric dimension that can be seen as an ecosystem for the organism, leading to an increasingly close morpho-functional correlation over time. Today, we understand that the genome of a species should not be viewed as a deterministic set of instructions that rigidly determines a specific phenotype. Instead, it can be likened to a recipe book with several possible interpretations (West-Eberhard, 2003). This does not imply that it can be translated in any arbitrary manner – otherwise, it would lose its informative role – but rather that it possesses a degree of virtuality resulting in various potential morphotypes. Hence the *intrinsic polyphenism of the genome*, wherein multiple phenotypes are already anticipated and correlated to specific environmental gradients (Simpson et al., 2011). This goes beyond the paratypical effect, whereby the environment affects embryo development: it involves the recognition that external gradients have a precise influence and are not mere random noise. For instance, in tortoise embryos, the temperature at which the eggs are exposed determines their conversion into one sex or another. This concept aligns with Eco-Devo theory, which suggests that inductions occur in response to factors such as nutrition, the presence of certain predators, contact with conspecifics, and more (Gilbert and Epel, 2015).

Considering these insights, it becomes evident that the notion of infiltration, which either astounds or alarms us in the context of cybernetic hybridisation, is actually only the explicit manifestation of a very old phenomenon, which previously remained concealed because the device was external to the body. However, even a simple chipped stone has had an impact on morpho-poiesis, influencing praxis, oculo-manual coordination, frontal lobe enhancement, mnestic systems, and anterior train muscles (Pettitt, 2022). Conversely, this very early technology led to a reduction in other organs that were replaced by it, resulting in functional decline within individuals due to reduced use and within the population due to natural selection. Crafting a chopper or a hand axe involves shaping the flint, but this activity shapes the body as well, enhancing certain functions while diminishing others. This influence occurs at both the individual and species level, acting on selective pressures (Lewontin, 1983). *Functional externalisation*, which involves transferring a bodily function to a device, reshapes the entire morpho-poietic and physiological systems, operating in three areas: i) the ontogenetic domain, where organs are exercised differently, influencing functional regulation, similar to the effects of bodybuilding; ii) the epigenetic domain, encompassing both hereditary factors like DNA methylation and environmental factors, which transform genomic polyphenism into specific phenotypes; and iii) the genetic domain, also known as the niche effect, which alters selective pressures on the population, thus affecting species-specific fitness and leading to the emergence of genetic mutations.

The most intriguing aspect of this influence pertains to the developmental process or phenotypic expression of the genotype during the embryo-foetal phase, recognising that any modification occurring at one point in the somatic configuration has ripple effects on other parts of the body. This is exemplified by the MYH16 mutation, which, by reducing the development of maxillary muscles during embryogenesis, alleviated tension on the bones of the neurocranium, promoting its gradual growth (Carroll, 2005). Once again, the fallacy lies in the deterministic conception of configurational factors (in this case, DNA-related), as the genome can give rise to multiple phenotypic variations due to a multitude of factors. The body is not an isolated entity that is solely based on an intrinsic recipe encoded in its DNA: rather, the genetic contents interact with the environment, which influences their phenotypic expression. We can view the genome as a recipe book of protein possibilities, but their actualisation relies on *external inducers* that are diverse and exert their influence in various ways: some are present in the developmental environment of the zygote, such as the maternal environment in the case of mammals, while others are inherited from the parents' lifestyle and are found in the cytoplasm of the zygote itself (Skinner et al., 2010).

It is evident that the technosphere also functions as a cofactor or ecological element capable of influencing the translation of the genome into a phenotype. For instance, plastic has the ability to release substances that, when ingested by the mother, can alter the mechanisms of sexual predisposition in the embryo (Thorson et al., 2021). A specific diet, lacking certain protein or vitamin components, which can be attributed to certain food production practices, can impact the development of the pelvis and result in growth abnormalities in the embryo (Tiffon, 2018). These effects are not confined to the embryo-foetal phase but have repercussions on the adult age, such as an increased predisposition to cardiovascular diseases (McKay and Mathers, 2011). Since the advent of Caesarean sections, there has been a 20% rise in cases of fetopelvic disproportion, indicating a mismatch between the size of the foetus at birth and the maternal pelvis (Cregan-Reid, 2018). The rapid advancement of devices in the *Homo sapiens* species has led to a significant reduction in brain size over the past 50,000 years (Yakoubian, 2021). So, in every sense, the technosphere acts as a niche that influences all stages of the morpho-functional development of the phenotype.

It is important to acknowledge that the genotypic makeup of a species is not solely determined by passive factors such as mutations, meiosis, and crossing over, as was once believed. Other factors, such as symbiosis interventions, genetic assimilation, phenotypic accommodation, and niche construction, can also influence the phylogenetic trajectory (Kauffman, 2019). These new factors of phylogenetic influence challenge the notion of individuals as passive entities or mere carriers of genes. Instead, each organism, through its individuality and plasticity, contributes to shaping the genetic characteristics of the population. This underscores the significance of the interaction with the environment in the phylogenetic trajectory. For instance, the niche construction theory demonstrates how an individual's actions on the environment can alter the menu of selective pressures: this involves hiding certain pressures, emphasising others, and initiating altogether new ones. As a result, the fitness value within the population undergoes changes. If a particular morpho-function was previously a crucial selection factor but later wanes in importance while another morpho-function takes on greater significance, this leads to a transformation in the species (Reader and Laland, 2003).

There is no doubt that the construction of technical apparatuses should be seen as a process of modifying the niche, transforming fitness value, and consequently, causing a metamorphosis in somatic conformation. The body undergoes reconfiguration due to the changing selective pressures acting on the population, as if the living environment of that particular species had undergone a change (Laland et al. 1999). Prior to the use of chipped stones and fire for cooking, for example, there was a selective pressure on jaw strength for cutting and tearing food. However, as time progressed, this selection factor gradually diminished, while new selective pressures emerged, focusing on individuals' practo-gnostic capacities, operational memory functions, and imaginative skills. Each device not only shields the body from a specific selection factor but also creates new selection events, leading to what can be described as a selective shift (Vatsiou et al., 2016). When discussing the influence of technology on the body, both as a structure and a process of functional regulation, we must not limit our analysis to the ontogenetic process alone. Its impact also occurs, albeit at different rates, within biological timescales.

However, I want to be very clear: the notion that the body is a fluid entity entirely dictated by external apparatuses is false. Of course, it is important to recognise that the body possesses a degree of plasticity, akin to the evolvability of a species (Colegrave and Collins, 2008). There is a spectrum of expressive possibilities that gradually decrease, with the boundaries of translation defined by constraints. That being said, emphasising plasticity is very different from suggesting that the body is entirely malleable, as that leads straight to environmental determinism. The device operates within the realm of these possibilities, leveraging the system's inherent plasticity and relying on its expressive redundancy – not seeking to fill a void or impose a form on an amorphous condition (Edelman, 1987). The device does *not* completely redefine the body, as some transhumanist theorists may suggest. Instead, it explores new paths within the existing range, taking into account the potentialities and constraints of the somatic dimension. We are not merely an assemblage of modular morpho-functions; we are an organism in which each component, albeit in varying ways, influences all others. For instance, our eyes engage in a precise anatomical and physiological correlation with our brain. Therefore, simply modifying our sensory inputs would not suffice to transform us into the *Umwelt* of another species.

The technologies that immerse us in virtual reality rely on the inherent *Gestalt laws* specific to our species; they operate through sensory illusions that align with our perceptual characteristics (MacNamara, 2017). As said, while technology certainly transforms the organism, it does so within a range of possibilities. For this reason, exotic sensory abilities, such as the echolocation of microbats, cannot simply be added to our repertoire, but must be translated into optical images. Similarly, while physiology can be modified in its values, it cannot be completely revolutionised, as it is governed by a complex system of functional interactions and recurrences that operate within specific gradients. The same considerations apply to the ethological characteristics of our species (Eibl-Eibesfeldt, 1989). As mammals, we exhibit strong parenthood; as primates, we have a playful and competitive nature; as apes, we have a strong inclination towards imitation; as collectors, we have a deep-seated impulse to collect and catalogue; and as unique artisans, we have an inherent drive to utilise objects and phenomena. Therefore, our existence as cyborgs is always realised within a distinct phylogenetic dimension. Any device we encounter inevitably interacts with this combination of tendencies, systemic interactions, functional structures, and motivational propensities.

The presence of inherent characteristics in every living being cannot be disregarded in the name of an ideology that advocates complete emancipation from nature. This arrogant and perilous dogma, which has thrived within humanist thinking, suggests that to be truly human, one must detach oneself from any natural heritage. Despite Darwin's insights, this viewpoint continues to exist in transhumanist proclamations (Moravec, 1998). On one hand, this perspective represents nature as passive, characterised solely by extensionality and devoid of any creative capacity, trapped in a monotonous cycle punctuated only by blind chance. On the other hand, it exalts the poetics of humanity, portraying it as entirely liberated and creative, the master of Being, emancipated from any laws governing the natural world. This oppositional dualism serves only to incessantly argue that humans are *ontologically unique* – certainly not a particular kind of animal, but rather the plastic expression of non-animality. On the contrary, what is needed is a complete paradigmatic shift, a departure from the dichotomies of humanism, and the inauguration of a *new philosophy of technology* (Franssen, 2009).

Recognising the influence of technology on the human condition entails transcending the notion of absolute self-determination. It involves acknowledging that, like any other organism, humans are interdependent with their environment, shaping and being shaped by it (Moczek, 2015). Understanding the continuity of physiological adaptation mechanisms and bringing together morphogenetic models that are functionally linked to the environment is a challenging task (Bateson, 2005). This difficulty arises from a cultural foundation that has historically separated humans from their interconnections with the environment, perceiving them as pure entities unaffected by external influences, and emphasising their self-regulating nature. However, comprehending what it means to be a cyborg requires embracing an attitude that recognises the inherent predispositions of the species and its intricate relationship with the environment (Dennett, 1996). Despite this, we still clash against the humanist perspective, asserting that devices are mere tools with minimal influence on humans. According to this view, the responsibility lies solely with human free will, tasked with utilising these tools effectively. On the other hand, some perspectives consider humans as mere vessels of possibilities, susceptible to any kind of modification and even complete replacement, while still maintaining a discernible essence (Wagner and Laubichler, 2004).

Both these approaches are mistaken. Human beings are not entirely free, nor can they transform into any entity whatsoever, but they do possess a degree of freedom and avenues for escape – albeit, of course, not from human nature itself. This freedom, indeed, is facilitated by the phylogenetic heritage inherent in our species (Clark, 2003). Thus, it is essential to gain a better understanding of the characteristics and potential of these innate elements when using the various technological devices that become available. Discussing cyborgs does not imply dissociating the human from the morphogenetic models present in nature; rather, it means focusing on the distinct features of the niche, specifically the significant role that technology plays in shaping the somatic definition of our species. This occurs through three primary processes:

1. The dependence effect, whereby each technological externalisation subsequently engenders a form of addiction that manifests itself on various levels (Marchesini, 2002).

2. The tension effect on the body's physiological equilibrium, leading to allostatic consequences that necessitate a repositioning of homeostasis (Piazza, 2019).

3. The domestication effect, which entails a systemic transformation of the body, resulting in diverse functional regressions such as neoteny (Montagu, 1956).

Therefore, it is imperative that we analyse the effects of technomediation while considering the inherent morphological, physiological, and behavioural characteristics of the human species.

The second aspect I would like to emphasise pertains to how the ethological inherencies of our species interact with the opportunities presented by new technologies. In essence, we need a *techno-ethology* that can establish connections between human predispositions and the possibilities opened up by emerging technologies. Despite the widespread notion of (self-attributed) human rationality, modern neurobiological studies caution us that our choices are predominantly driven by affective impulses (Panksepp and Biven, 2012). It is our emotions, motivations, innate inclinations, *feelings*, and *passions* that interpret and are influenced by the available devices. This is evident throughout history; indeed, psychological research confirms that decisions invariably originate from affective predispositions. Therefore, we must not disregard the inherent competitiveness within our species and harbour illusions of some idyllic future where humanity is pacified solely by technology (Wilson, 1978).

I am not advocating a technophobic stance, but rather a critical examination of social and techno-cultural transformations, grounded in reality. There is no point in claiming that the key factor is *how we choose to utilise a particular technology*, as this assertion rests on two far-from-obvious assumptions: (i) that technology does not bring about changes in the human condition, encompassing values, orientations, objectives, consciousness, and especially epistemology; (ii) that the usage of the device follows rationality and free will, is unaffected by the affective influences inherent to humans, and is not subject to the logics of power, segregation, and inequality that characterise different societies, even those with democratic foundations. History has taught us that, particularly in its initial stages, each device tends to *widen* the gap between the powerful elite and the rest of the population (Warschauer, 2004). It is no coincidence that the technological acceleration of the 20th century has exacerbated this divide, with 1% of the population possessing more wealth than the remaining 99% and 10% of the population consuming 90% of the available resources. On the other hand, we cannot ignore the vulnerabilities that human nature has displayed regarding the effects of technology, such as addiction and euphoria. Every device disrupts physiological balance, and it is crucial to acknowledge this fact, not to reject innovation, but to implement countermeasures and develop a more resilient *psycho-immune system* (Esposito, 2020). In general, the danger does not stem directly from the technology itself, but from the impact it has on the entirety of the human condition. With each innovation, our very configuration changes, and we are no longer what we were before.

We can envision the development of a device as the establishment of a new symbiosis that, akin to natural symbiotic relationships (Gilbert and Teglborg, 2023), prompts us to: 1) prune certain characteristics that become redundant or hinder the complete realisation of the new functional alliance; 2) foster the development of other functions that complement the symbiotic interaction. In natural associations, be they true symbioses, mutualisms, or even parasitisms, organisms undergo organ loss – e.g., insects losing their legs or wings – or a reduction in certain enzymatic properties, as these functions are performed by the symbiotic organism. Today we know that many symbioses even involve a significant reduction in specific genes (McCutcheon and Moran, 2011): for instance, the fig wasp *Ceratosolen solmsi*, which spends most of its life inside the fig, has lost a significant portion of the genes responsible for environmental perception (Xiao et al., 2013). In addition to this, it is crucial to recognize that a symbiosis can only be fully realised through a process of functional accommodation, which entails complementarity (Rosenberg et al., 2007). When evaluating a device, it is therefore not enough to solely consider its usage – whether using it in a particular way makes it good or bad. Rather, it is paramount to first comprehend the impact it will have on the human condition, considering the key factors that define it (Gallagher and Zahavi, 2008).

We are aware that the incorporation of a technical apparatus leads to an ontological deviation in the human being and can give rise to problems for which we may not be sufficiently prepared. It is understandable, then, that some advocate for caution, which should not be interpreted as a rejection of technology, but rather as a reflection on the potential evolutions that a particular technology may bring to human life. An example of this is the request for a six-month moratorium made by a group of scientists in spring 2023 regarding the development of highly advanced artificial intelligence devices like Gpt-4. The purpose of such a moratorium is to assess the impact of these technologies on human life (Narayan et al., 2023). Failing to consider this factor would put us at risk of misunderstanding the ontopoietic space being shaped, which cannot be solely rooted in objective performance qualities or the definition of control over usage. Acknowledging the concept of exposure means recognising the interactive-emergent effect of the technical apparatus as it interacts with the characteristics of the human condition, particularly its specific phylogenetic heritage (Di Francesco and Marraffa, 2009).

As said, when a new technology emerges, it disrupts the entire ecological landscape of human existence, bringing about transformations that are influenced by both the specific characteristics of the technology itself and the human condition, encompassing both innate heritage and the accumulated technical-cultural dimension. Acknowledging that humans have always been shaped by the tools they use does not imply denying the presence of a distinct species identity that incorporates and integrates these tools within its own ecological context (Allenby and Sarewitz, 2011). Rather, disregarding this would mean attempting to dissolve human nature to the extent of considering it entirely determined by the external environment. So, by moving beyond the binary logic of determinism or the simple juxtaposition of "innate and learned," we will come to realise that it is entirely possible to uphold a complex and specialised human nature while also recognising its remarkable adaptability to external influences (Descola, 2005). It is precisely the ethological characteristics of the human being that are affected in the process of incorporating technology: in some ways, they have facilitated hybridisation and the emergence of the cyborg, while in other ways they have suffered from it, particularly in terms of acquired dependencies (Amoretti and Manzotti, 2012).

On the other hand, it is the morpho-physiological condition of humans – characterised by neurobiological redundancy, a predominance of technical and symbolic intelligence, neonatal immaturity coupled with strong parental care, motivational proactivity, flexibility in body structure, and a highly developed social dimension – that has fostered a pronounced and self-perpetuating inclination toward hybridisation within our species (Wrangham, 2019). In fact, it would be more appropriate to call humans *Homo faber* rather than *sapiens*. In fact, wisdom doesn't appear to be our strongest attribute, whereas we do represent the culmination of a phylogenetic tree marked by a strong propensity to use and construct tools. The act of construction entails learning how to handle stone, fire, skins, metals, bone, clay, and eventually electric circuits. This acquisition of such techniques goes hand in hand with the development of skills such as body movement organisation, envisioning the final form of the tool, refining eye-hand coordination, and utilising other auxiliary tools to manipulate matter. In short, the process of construction goes beyond the mere act of making a device; it means modelling the body and the mind (Stout et al., 2008). The tool is deeply ingrained in the neural configurations that connect the eye to the hand or enable the recognition of form within amorphous structures and the intricate interplay between muscle movements and postures specific to a given activity. Construction is a recursive process with multiple, often unpredictable effects, sometimes self-catalytic and at other times homeostatic, frequently entailing threshold effects and always leading to the emergence of new attributes and capabilities.

Threshold effects refer to systemic phenomena characterised by reciprocal interactions between factors. Up to a certain value, known as the system's threshold, regulation occurs in a homeodynamic manner, driven by feedback- mechanisms that maintain the system within a homeostatic dimension. However, beyond this threshold, a different type of autocatalytic regulation, referred to as feedback+, comes into play, leading the system nonlinearly into another functional dimension (Marchesini, 1996). Many natural homeostatic systems exhibit threshold system characteristics. For example, consider the carbon cycle, which balances the release and absorption of carbon dioxide by plants and animals. As long as the CO2 levels stay below a certain threshold, the system remains stable. However, if the CO2 levels rise, it can trigger a greenhouse effect, leading to overheating and the destruction of plants and CO2-fixing plankton. This disruption not only breaks down the homeostatic mechanism but also contributes to increased CO2 emissions (Libes, 2015). The strength of threshold systems thus lies in their fragility, highlighting how gradual and hidden changes can suddenly result in significant transformations. Evaluating change from a systemic perspective also involves recognising that interventions should not be limited to isolated points but consider the broader systemic context (Johnson, 2001).

Constructing a device goes beyond acquiring technical knowledge and skills; it also involves developing metacognitive qualities that are inherent to the process. These include precision, diligence, attention, concentration, organisation of materials, order, sequentiality, and collaborative teamwork (Metcalfe and Shimura, 1994). The level of knowledge transmission is closely tied to factors like increased life expectancy, allowing for the presence of elderly individuals with advanced proficiency in device usage, as well as the extension of childhood and the capacity for learning. The question of what came first, whether the metamorphosis of developmental times (as shown by comparing fossil records of *Australopithecus africanus*, *Homo ergaster* and *Homo sapiens*) or technological advancement, is misguided as it assumes a linear causal relationship. Understanding the impact of technology on the body requires a systemic perspective on the somatic dimension, avoiding the narrow focus on a single function. Incidentally, it's worth noting that genes exhibit pleiotropic effects, producing diverse outcomes in different parts of the body (Paaby and Rockman, 2016). However, the organism itself operates as a system, meaning that every transformation has far-reaching consequences, and any modification sets off recursive effects, triggering feedback mechanisms.

As the developmental timeline of a species changes, it inevitably impacts the overall social structure and the size of affiliative groups, subsequently affecting the cognitive structures responsible for managing internal coordination practices, as illustrated by Dunbar's number (Moffett, 2019). The adoption of a device not only influences practical and cognitive capacities but also impacts short-term or working memory functions. This includes prospective memory, involved in sequencing, and buffer memory, responsible for integrating relevant information. These changes alter the temporal perception of the present moment and have implications for how individuals interact with their surroundings, including their ability to selectively attend to specific stimuli and focus on particular aspects of their phenomenological experience.

In *Homo sapiens*, it is well-established that the increase in device production has led to progressive modifications in certain brain structures. One piece of evidence for this encephalic reorganisation can be observed in the emergence of a globe-shaped skull, which is associated with the expansion of the parietal lobes involved in working memory. Thus, technomediation has an undeniable effect on individual ontogeny, but it also results in morpho-functional refinements within the biological timescales of phylogeny. While it is true that humans have always constructed tools, it is equally undeniable that these tools have played a role in shaping our species as well. In this sense, we can envision ourselves as both the creators and products of technology – we are the other side of the mirror that reflects both the natural world and the realm of technology, as eloquently depicted by Konrad Lorenz's image.

It is essential, therefore, to develop a new awareness of this symbiotic relationship in order to cultivate a different philosophy of *techne* (Irrgang, 2021), which transcends both the self-sufficient conception of human nature and the notion of its assumed fluidity. Indeed, it is a mistake to believe that humans can fully reinvent themselves through anthropo-poiesis. Instead, we should embrace the model of the cyborg – an individual who recognises the hybrid nature of their existence, accepting the device as an integral part of their ontological dimension that does not deny but expresses their inherent human identity (Greguric, 2021). This awareness entails a critical perspective that acknowledges the mechanisms of dependence and the exposure technology entails. We have always been cyborgs; we simply were unaware of it, clouded by humanist ideology and the subtle nature of the effects of technology. However, it is undeniable that the very act of constructing a device involves physical and cognitive exercise, shaping the individual in ways that complement the skills required for the task at hand.

The adoption of a technology, in fact, necessitates the acquisition of *technai*, specialised skills of the body and mind that become ingrained within the individual, influencing not only their performance in specific tasks but also their self-efficacy, perception, perspective, sense of time, and overall mentality. Indeed, these *technai* permeate various aspects of one's life, extending into dreams, ambitions, desires, values, and the very definition of virtues. *A new kind of individual emerges when tools, and particularly their pervasive presence in daily life, reach a critical threshold* (Pellegrino, 2006). This gives rise to a condition where virtually every activity is the outcome of technomediation. Work involves employing devices to create new technical intermediaries, whether it be a composition, a piece of clothing, a surgical procedure, or the construction of a road. In every human endeavour, the devices used become extensions of the body, shaping the entire somatic dimension and guiding developments and orientations in morpho-poiesis. It is important to recognise that each activity involves the allocation of limited growth resources – every developmental advancement implies a corresponding regression. Ultimately, there is a profound correlation between the physical form of the body and what shapes action (Sakamoto et al., 2003).

If we compare various aspects of different professions, such as the manual grip strength of a worker versus that of a writer, the spatial memory of a London cabby (prior to navigation assistants) versus that of an average person, the thickness of a farmer's skin, or the olfactory capacity of a perfumer, we quickly realise how work becomes fully embodied in those who perform it (Maguire et al. 2000). On the other hand, it is also evident that human work cannot be disentangled from the tools and equipment that enable certain activities. A craftsman becomes one with their devices, perceiving them as extensions of their own body and integrating them into the parietal cortex. The same goes for an artist, who not only shapes their studio based on their preferences but is also influenced by it in their creative inspiration. In each of these activities, there are bodily incorporations – motor and sensory translations – similar to learning how to jump, dance, run, or climb. A craftsman –like an artist, or surgeon – has strengthened specific muscles and neuronal assemblies required for their work, operating through distinctive flow diagrams. The device is no different from a craftsman's hands; it becomes an integral part of them, enabling their perception of the hardness, malleability, smoothness, or porosity of the material they are working on.

Ultimately, the relationship between humans and technology has always been more than one of mere use – it is a symbiotic connection that inevitably gives rise to dependencies. Yet, it is not always easy to recognise these dependencies, especially once they have become ingrained. Thus, we tend to view technomediation as an external crutch or compensation rather than an intrinsic part of our being. This misconception arises from a lack of attention to bodily changes and the biases that view the skin as an impermeable barrier preserving the purity of human essence. However, with the advent of somato-technological hybrids, something within us has shifted. Our perception of the relationship with technology has changed, and a new anthropology has emerged, no longer rooted in the notions of purity and self-sufficiency, but in hybridisation. We are only at the beginning of a complex journey that necessitates a reevaluation of categories and boundaries – a transitional period in which familiar certainties vanish before new coordinates have solidified. It is understandable, therefore, that we feel overwhelmed by uncertainty, which leads to both unwarranted enthusiasm and groundless fears, as we struggle to define the emerging context. What is certain is that this metamorphosis is irreversible and will inevitably reshape our understanding of the human condition.

6.2 Behind techno-addictions

Technology has a way of ensnaring us in a lifestyle that swiftly leads to a genuine addiction. This first form of techno-addiction stems from the capabilities and conveniences offered by various devices, expanding our realm of operation and simplifying certain tasks (Mitchell, 2019). However, this addiction is not merely superficial, limited to habits and preferences; it permeates the very physiology of our bodies (Kringelbach and Berridge, 2009). Indeed, the body becomes intimately connected to the device, regulating the metabolism of its organs in the short term, modifying the phenotypic expression of the genome in the medium term, and even influencing the genotypic structure through the reconfiguration of selective pressures and fitness value in the long term (Nussey et al., 2005). Nevertheless, what we primarily experience is: 1) a perception dependency, i.e., the habit of relying on the device's functionalities; and 2) a short-term dependency, i.e., the regulatory alterations of various bodily functions. In essence, these two forms of dependency represent a novel way of experiencing the body through a technological apparatus: an irreversible process that inaugurates a new and unprecedented state of being for humans.

Habit addictions cannot be attributed solely to the specific functionalities of a device, as they encompass the broader impact it has on the entire technological landscape and on various aspects of people's life (Scarmeas and Stern, 2003). However, these addictions often stem from the services provided by the device, which simplifies certain tasks, expands possibilities, facilitates goal achievement, conserves resources, and enhances results. This form of addiction can be characterised as a dependency on performance, as acquiring a device that raises efficiency and effectiveness levels inevitably binds one to that particular technology (Sigman, 2009). It is important to note that when we talk about performance improvement, this is rarely a purely functional evaluation, as social factors strongly influence this assessment. The significance of a given device is tied to the social validation it offers individuals, particularly in competitive contexts but often in affiliative ones as well. It is this social affirmation that makes a certain level of performance desirable. Therefore, it would be erroneous to seek a strictly utilitarian understanding of techno-addiction, as it is frequently driven by the need to keep pace with society as a whole and the fear of falling behind (Dossey, 2014).

A specific form of dependency on performance can be attributed to services that enhance individual autonomy and the ability to navigate various situations. An example of this is the influence of digital assistants, omnipresent tools that simplify our lives by eliminating costly and cumbersome intermediaries. Indeed, these assistants go beyond mere convenience; they anticipate our needs based on knowledge about our preferences and orientations (Hoy, 2018). This addiction can be seen in phenomena like *digiphrenia*, which refers to the tendency to exist ubiquitously through simultaneous self-presentation, or multitasking. The desire to stay connected at all times with friends, acquaintances, and work contacts is also part of the pervasive connectivity we are now accustomed to. We rely on satellite navigation when we travel, our contacts are stored in our mobile phones and no longer need to be memorised, video calls are effortlessly accessible, and information can be retrieved instantaneously (Skues et al., 2012). What has fundamentally changed is our concept of presence, which no longer pertains solely to the here and now, but extends to a state of temporal and spatial expansion – what we could describe as *co-presence*. When a certain performance becomes customary, it is often taken for granted, as is the case with the concept of onlife – the notion of being constantly connected to the network. This condition further underscores the subtle and insidious nature of addiction (Tamir and Mitchell, 2012).

Continuing with the discussion on performance, we must also consider the chrono-dimensional impact of technology and its influence on our perception of time. Technology has the ability to shape the temporal structure of processes to which we become accustomed: today, we expect to download movies or files in a time frame that would have been considered utopian just a decade ago. Yet, we now find ourselves becoming impatient if the device takes too long to complete the task. Indeed, the introduction of innovative devices often leads to an acceleration of our experiences, which, over time, becomes ingrained as the rhythm of our lives.

Similar to how the sun resets our biological clock's circadian rhythm, technological devices have the power to alter our perception of time, as is already evident a few decades after the digital revolution. Our temporality has undergone a transformation, and we are now compelled to synchronise with machine time. Time seems to speed up due to the constant influx of images, the pace of films, the rapidity of requests in our daily practices, and the expectation of receiving immediate responses. Individuals also feel as if time is slipping through their hands, with opportunities exceeding the available time. The need to always be connected and up-to-date intensifies our "chronophagia" (time-consuming), and this phenomenon is not solely due to technological innovation as such but specifically to the mastery of increasingly complex devices with shorter obsolescence periods. However, the drawback is that haste erodes our capacity for reflection. The constant consumption of novelty predisposes us to boredom and irritation, while living in a continuous stream of stimuli causes our attention to wander (Appelbaum et al., 2013).

Another significant aspect to consider is the gratification effect and, to some extent, the *hedonic trap* that often lurks behind technological usage, which represents the strongest form of addiction. The initial form of pleasure can be attributed to performance, where the mechanisms of self-efficacy come into play: a device can satisfy the subject's performative narcissism, making them feel stronger and more capable of facing the challenges of the world. Additionally, there is the pleasure derived from the gratification system, where a dopamine-mediated connection is established between the desired outcome and the means that enable it (Berridge, 2007). However, it is crucial to provide some clarification at this point; otherwise, we may be led to believe that it is a simple transition from one form of gratification to another, without any adverse effects. The digital experience holds an undeniable surrogate value: instead of embarking on physical journeys, we immerse ourselves in virtual realities, and instead of engaging in real adventures, we rely on avatars, among other examples (Palmer, 2007). What changes is the level of physical participation in these experiences. It is widely recognised that a decrease in physical activity leads to a decline in muscle endorphin and serotonin levels, resulting in a reduction of the satisfaction gradient and mood balance. This, in turn, can lead to increased feelings of yearning and a tendency to repeat the experience. The surrogate action is never a true replacement because it fails to generate the physiological responses that are present in actual activities. Consequently, a compulsive mechanism is triggered, akin to the repetitive behaviour exhibited by animals in captivity as they pace back and forth in front of bars. In other words, the decline in proactivity and the reliance on surrogates give rise to dependency (Owen et al., 2001).

Not to mention, the reduction in physical movement and limited exposure to sunlight have been contributing to an increase in a wide range of pathologies. The cardiovascular system is one of the first to be affected, as physical activity is the foundation of its health, influencing factors such as blood pressure regulation, oxygenation, vascular cleansing, and the conditioning of the heart muscle. It is no coincidence that diseases like stroke and myocardial infarction have become leading causes of mortality (Rutledge et al., 2008). The immune system has also become compromised, as its efficiency relies on the muscle pump. We have observed a rising trend in immune-related disorders, including allergies and autoimmune diseases, which, even if not solely attributable to physical activity and connection with nature, do exhibit worsened outcomes due to people's increasingly sedentary lifestyles. Inadequate exposure to sunlight leads to disruptions in the sleep-wake cycle and a deficiency of vitamin D, the latter being crucial for proper bone mineralisation. This deficiency gives rise to various disorders, such as rickets, osteomalacia, muscular pain, hyperparathyroidism, certain skin abnormalities, and immune deficiencies (Cumming and Vandewater, 2007).

Another form of addiction arises from the user engagement mechanisms implemented in many digital environments through features such as alerts or notifications that aim to capture the individual's attention and encourage continued usage. Platforms strive to attract and retain their audience as it enhances their value. Thus, the use of "mementos" is encouraged whenever users appear to be losing interest or neglecting their regular interaction with the application (Leonhard, 2016). As Günther Anders predicted, this perpetuates a state of constant need oscillating between production and consumption, resulting in an ever-increasing dependency. Anders referred to the process of familiarisation, which normalises and routinises the dynamics of self-expropriation, effectively concealing alienation in both its causes and symptoms. Humans perceive themselves as shepherds of the machine, deceived into believing they act spontaneously while the real decision-making power remains hidden. This state, in which reminders become services readily available to individuals, serves to demonstrate once more the power of technology, for it does not need to reveal itself explicitly – the absence of manifestation makes obedience appear natural, creating an illusory sense of freedom. Every morning, upon awakening, one is already surrounded by seductive sirens, ready to entice one's attention (Douglas, 2020).

Digital technologies create addictive patterns by requiring users to memorise scripts and conform to specific practices, effectively shaping their cognitive processes and brainframes. By providing pre-packaged content tailored to individual preferences, users are enticed to remain engaged with the media, resulting in a form of consumer addiction. Certain platforms, such as Google, exemplify this with their high degree of stickiness by offering a wide range of available services, in this case email, translation, maps, and search engine. This phenomenon is often referred to as digital bulimia, a compulsive addiction that drives individuals to incessantly engage with digital technologies, even during sleep, while driving, during important conferences, or in other high-stake contexts (Alborghetti, 2022). However, this addiction extends beyond mere habit. It is intricately connected to the neurophysiology of dopamine, where two processes occur: (i) the expectation system becomes increasingly appetitive due to a lack of fulfilling experiences and variable reinforcement mechanisms, and (ii) the entire affective seeking system becomes tightly channelled into the virtual realm, further exacerbating dependency.

Dependency can also be attributed to the profound epistemological significance that technologies have always held in human lives, in terms of both imagination and the desire to explore uncharted territories. Our innate disposition as exploratory beings, driven by curiosity and the pursuit of expanding our horizons, has a close connection to the technologies prevalent in a specific historical period, culture, or society (Sigman, 2009). The term "technoscience" is used to emphasise that the advent of a particular device should be understood not only in terms of its practical applications but also in terms of epistemological redefinition. Scientific knowledge has also advanced through the technologies available in a given era, granting human beings access to phenomena that were otherwise inaccessible. Science, as a collective endeavour embedded in a specific historical context, undeniably relies on the development of devices (Olsen and Christiansen, 2009). In all its forms, a device serves as a medium to transcend known boundaries and thus embodies the spirit of human ingenuity. Each technological tool not only enables but also defines the possibilities available and opens up unique dimensions of the human experience.

A similar argument applies to the various cultural aspects that establish social belonging, such as identification and self-recognition within a specific society, encompassing customs, habits, styles, values, rituals, and taboos (Prensky, 2001). The need for belonging is a fundamental aspect of social behaviour, where individuals strive to keep up with the scientific knowledge and cultural expressions of their time. As social animals, we are accustomed to organising into groups based on an "us vs them" dynamic, relying on identity apparatuses. When we consider devices not only as useful tools but also as elements of identity recognition, we quickly realise the level of dependence that technical apparatuses generate. Digital technologies have intensified individuals' dependency on these mechanisms of belonging by deconstructing their identities into profiles, lacking a solid foundation and transforming them into a collection of atomised and potentially disconnected characteristics (Floridi, 2014). Algorithmic profiling focuses not on the person, but on their behaviour, including orientations, choices, habits, consumption patterns, and lifestyle. Profiling does not require internal consistency; instead, it breaks down the individual into constellations of traits that produce predictable phenomena. Consequently, personal identity becomes fragmented and contradictory, further increasing the individual's need for superficial attachment to current trends. This dependence is evident in generational styles (Warner and Sandberg, 2007).

One significant aspect of belonging is affiliation, where individuals are recognised as similar and thus are able to be part of a group. This fuels the human tendency to conform, the need to align with the prevailing trends and the given *Zeitgeist*. Breaking free from this process of conformity is challenging, and those who dare to do so often face steep consequences, including rejection, discrimination, stigmatisation, or the risk of becoming a scapegoat (Haddon, 2000). Consequently, following the societal current has become the primary means of feeling fully integrated, and technologies play a pivotal role in this process. Indeed, they are not merely performative tools but, above all, social operators that facilitate participation and create conditions for dependence. It is undeniable that, for today's youth, being deprived of a smartphone or tablet equates to being socially isolated. Without these devices, individuals not only miss out on participating in group rituals but also face a higher risk of bullying due to their perceived difference. Even educational institutions, starting from primary schools, heavily rely on digital platforms for school records and teaching activities. This reliance has been further exacerbated by the Sars2 pandemic, which led to lockdown measures and the widespread adoption of online lessons and webinars, effectively establishing digital classrooms.

The digital divide has been extensively studied by sociologists, who describe it as a form of digital marginalisation or a gap between those who have access to technology and those who are partially or completely excluded from it (Suler, 2004). Various factors contribute to this exclusion, including economic conditions, education level, age, and cultural background. The digital divide encompasses both the possession of state-of-the-art technologies and the skills to effectively utilise online information. A digital outsider is not only someone who lacks the necessary resources to fully leverage the potential of the Internet but also someone who lacks the knowledge and skills to navigate digital media. It is worth noting that an increasing number of professional and economic opportunities have shifted to the online realm (Burke et al., 2010). Indeed, the rise of influencers on platforms such as Instagram and TikTok exemplifies how proficiency in these digital mediums can lead to social and professional recognition. These phenomena are self-perpetuating, as the more an individual gains popularity through likes and shares, the more their content will spread virally across the Web. Two types of addiction arise from these dynamics: firstly, young people are increasingly driven to seek their fortunes within the new digital Eldorado, aspiring to make a name for themselves and becoming increasingly attracted to and dependent on these technologies. Secondly, influencers wield a growing influence over young people, creating new addictions in terms of followers adopting them as role models, copying their lifestyles and behaviours.

The social dependence on digital media far surpasses that of earlier analogue devices, although the media revolution of the 20th century, including radio, television, and cinema, had already shaped the need for technologically-mediated exposure and integration (Eco, 1965). It is therefore crucial to consider the social dynamics that unfold within the complex interplay of conformity and individuation, processes that are always situated within a technomediated environment. Additionally, two other pivotal aspects of social participation – collaboration and competition – must be taken into account as they are deeply intertwined with existing devices. In fact, in order to engage in collaboration with others, individuals must possess the technologies that are commonly used by others, even if it's just to enter into a cooperative system-environment (Konrath et al., 2011). Competition plays an even more significant role in dependence, as it necessitates a constant pursuit of improvement, the enhancement of services and capabilities, and a fervent consumerism of technological innovations. It is not coincidental that companies selling digital technologies and services constantly release new versions of their products, finding a user base eager to embrace this drive for consumption and passively accept the rapid obsolescence of devices. The competitive motivation to keep up with new technologies stands as the primary driving force in this dynamic, and this need for affirmation breeds addiction (Loader, 2007).

The primary form of dependence revolves around experience, which is closely linked to the sentimental education that individuals undergo during their early childhood years. Experience, in essence, represents a sense of rootedness and connection to the defining elements of one's formative environment. Put simply, our ontogenetic journey inherently encompasses various forms of dependence as we become intricately intertwined with the ways of life, styles, and habits we developed in our youth. Experience also reflects the level of emotional coherence we feel when encountering a particular situation, sometimes leading to feelings of confusion, vulnerability, and discomfort, or conversely, a sense of fulfilment, well-being, and suitability (Damasio et al., 1996). Throughout development, we acquire different forms of dependence by grappling with the influences that affected us during our childhood. Familiarity, for instance, relates to the attachment process, which encompasses not only emotional connections like having a secure base but also involves boundary elements, such as transitional objects (Winnicott, 1971). This implies that anything with which we have formed an emotional relationship during our developmental years exerts a bond based on oxytocin, thereby evoking feelings of security or nostalgia.

This is, in fact, the *dependency on objects and devices from one's childhood*, which influences how individuals interact with technological devices (Purcell et al., 2012). When comparing different generations today, we observe the following: (i) Older individuals often feel marginalised by new technologies and may display expressions of rejection. However, they still experience dependencies as they rely on others for carrying out everyday tasks. (ii) Middle-aged individuals attempt to navigate new technologies but may experience stress and feelings of ineffectiveness, which amplify their dependence on the technological medium. They often lack an affective bond with these devices, experiencing a sense of estrangement. (iii) Younger people not only possess a high level of proficiency in using these devices but have also formed an emotional attachment to them, considering them as transitional objects from their childhood. They have an affective relationship with these devices (Konrath et al., 2011). Indeed, younger individuals exhibit both mental agility in navigating new media and an emotional dependence on them. As said, what we experience during our first ten years of life is our sentimental education, forging bonds that extend beyond mere device usage. Various factors mediate these bonds, with oxytocin, a neuromodulator associated with emotional attachment, playing a significant role. Consequently, young people demonstrate an addiction to digital media that transcends performance-related aspects, as it involves their own emotions (Wang et al. 2015).

Experience is influenced by a combination of character variables and by the sentimental education one has received, shaping how individuals navigate specific situations (Tosun and Lajunen, 2009). Additionally, experience is contingent upon interface availability, namely the challenges or opportunities inherent in the situation. These factors include: (i) Access to information necessary for staying up to date, consulting media, or conducting research. (ii) Tools available for addressing specific needs and the level of proficiency in utilising them to achieve desired outcomes. (iii) Contact with one's affiliation referents or participation in social circles. (iv) Resources that facilitate the realisation of dreams and aspirations, enabling individuals to overcome constraints and pursue their life projects. (v) Means to maintain physical and mental well-being, recover from debilitating illnesses, and prevent or alleviate physical and psychological distress. (vi) Opportunities for exploration, freedom, and travel, which allow for the discovery of new existential dimensions, exposure to different territories, and acquisition of new experiences. (vii) Devices that enhance knowledge acquisition, facilitate investigation of reality, and enable the testing and refinement of scientific theories. (viii) Access to leisure and entertainment options that provide exposure to diverse artistic expressions, foster imagination, and alleviate boredom, sadness, and the limitations of everyday life. Technology plays a significant role in all these aspects.

In many ways, we swiftly adapt to the pleasures that arise from the widespread availability brought by technology, leading each device and innovation to be quickly integrated into our daily lives. With technological assistance, we feel empowered and experience a heightened sense of presence, as if everything were more concentrated, vibrant, rhythmic, and, most importantly, within our grasp. It feels as though our channels of experience have expanded their capacity, enriching our lives in the process (Anderson and Rainie, 2012). Technology also has a calming effect, shielding us from disturbances and acting as a comforting nest or a safe haven that protects us from the harshness of the world. However, technology also surprises us, generating an adrenaline-like effect as we marvel at the possibilities it unveils. This ever-evolving aspect of technology, in a perpetual state of advancement, instils pride in us as its creators and keeps us in a perpetual state of awe (Clasen et al., 2019). It is important to note that adrenaline itself can be addictive, as experienced by those who engage in extreme sports. Therefore, the mere idea of depriving ourselves of the means that have facilitated this qualitative leap in our experiences feels like a regression to a state of existential vulnerability. Experience becomes a constant comparison between the capabilities of the body in its raw form and what is made possible through technological support. The somatic dimension is inevitably perceived as deficient because we have grown accustomed to an alternative way of life, one shaped by technology.

6.3 Addiction and techno-stress

The relationship between the body and technology is further manifested through mechanisms of dependency that involve actual modifications in somatic physiology. These alterations can sometimes be reversible, but more often than not, they tend to be permanent (Howard-Jones, 2011). Indeed, even if reversibility were possible, it would require suspending the regulatory mechanism imposed by the device, which is unlikely to occur. Consequently, the dependency is more likely to manifest as a reconfiguration of technomediated homeostasis, or the process may progressively worsen, necessitating increased reliance on technological mediation. An illustrative example of this phenomenon is seen in the widespread drug abuse prevalent in contemporary society. Individuals often find themselves needing to increase the dosage of a substance in order to maintain functionality, as the body becomes habituated and decreases its production of endogenous homeostatic factors (Tinsley et al., 1994). These regulatory processes are mediated by technological interventions that engender addiction, as somatic physiology loses its functional autonomy and comes to necessitate the auxiliary intervention of the device.

The first crucial aspect to consider is the phenomenon of performative dishabits and the down-regulation that progressively fosters the body's dependence on technological support. This was previously discussed in relation to the varied development of different bodily apparatuses based on their stress levels. For instance, Wolff's law elucidates how bones respond to load stress by becoming stronger through repeated exposure, while Davis's law explains how ligaments and soft tissues lengthen or thicken when subjected to moderate but prolonged tension (Cregan-Reid, 2018). Conversely, these tissues undergo a reverse process of material reduction when constantly relaxed. The law of use and disuse also applies to the endocrine, immune, and nervous systems (as exemplified by the theory of neural Darwinism), and even influences the intestinal microflora and metabolism. Therefore, each time the performance load on a bodily apparatus diminishes due to functional replacement by a device, a process of somatic performance reduction is initiated, inevitably leading to dependence. The widespread use of antibiotics or psychotropic drugs, the excessive reliance on devices for cognitive functions like calculation or memory, the decline in physical activity in terms of musculoskeletal exercise, and the prevalence of processed or minced food have collectively contributed to our increasing dependence on technologies and corrective devices.

Another form of addiction arises from the phenomenon of over-stimulation of a receptor or function induced by a device or its consequences, known as up-regulation. This mechanism involves the progressive increase in membrane receptors, leading to heightened cell sensitivity and resulting in metabolic overproduction. Many drugs act as antagonists, preventing neurotransmitters from binding to receptors and causing an increase in postsynaptic receptors. When these drugs are discontinued, a rebound effect occurs due to the higher receptor numbers. Up-regulation can also affect the endocrine system, altering hormone response through exposure to toxins, or impact the immune system by inducing cytokine production as a result of repeated inflammation or system inactivity. For instance, liver cell function can be affected by up-regulation upon exposure to xenobiotic molecules present in certain pollutants like dioxin, triggering an increase in cytochrome enzyme production. Some forms of up-regulation are associated with muscular activity: for example, certain exercise practices can heighten cellular sensitivity to insulin (Hennige et al., 2003). Additionally, up-regulation can affect constantly activated neural connections, leading to myelin sheath thickening, increased postsynaptic dendrites, and heightened sensitivity to neurotransmitters (Carlson, 2013).

A third form of addiction can be attributed to the mechanism of modulative conformation, which refers to the configuration of the system induced by the device. One example of this is the dopaminergic system, which regulates seeking, expectation, and reward, and can be hacked generating addiction. Dopamine release is not solely tied to reward but is also influenced by the uncertainty or unpredictability of outcomes and the element of surprise when the result surpasses expectations (Linnet et al., 2010). Many addictions, including gambling addiction, rely on variable reinforcement mechanisms and the jackpot effect. Numerous digital games exploit these mechanisms, and even the outcomes of social media activities such as likes and shares can impact these neurobiological processes. The dopaminergic effect not only fosters addiction but also intensifies cravings and the individual's engagement within a particular context. It is now understood that dopaminergic pathways are remodelled by one's experiences and become oriented towards the source, in this case, digital media (Koepp et al., 1998). Dependency through modulative conformation affects all systems that incorporate the device and become reliant on it, as they integrate it into the physiological chain sequence.

The fourth form of addiction relates to functional conformation, wherein the inclination to perform a specific task through a device not only provides practical advantages – making the performance easier, more effective, more extensive, and readily available – but also instils an organisational configuration within the body. This conformation establishes particular synergies between apparatuses and functional sequences (Sigman, 2009). As previously mentioned, we tend to become accustomed to new devices, which lead us to quickly conceive our performance within their operational framework, relying on their efficiency, efficacy, or specific outcome. However, the emergence of hybrid performance also engenders a distinct functional conformation of the somatic apparatus. Take, for example, the act of driving a car. During this process, we develop a synchrony between different body parts, constructing an internalised procedural dimension. As happens with riding a bicycle, playing ping-pong, or assembling a puzzle, we prepare the body to execute the intended function and, in a sense, *conform it to incorporate the instrument* as an integral part of the organism's continuum. Thus, dependence is formed on a particular mode of performance, wherein the body becomes reliant on the device as an essential component of its functional capabilities (Smith, 2005).

In this sense, the specific performative variation of the body becomes complementary to a device to produce a hybrid performance that is progressively assimilated as an integral part of the somatic dimension. This is why the motorcyclist feels like a centaur and a prosthesis is fully introjected as an additional organ. This produces the explanatory error whereby we believe that it was the insufficiency of the body that prompted techno-poiesis and not the other way around. In fact, it was the advent of technology and thus of technomediated performance that brought out – as a consequence and not as a cause – a lack in the body. In this light, one can clearly see the sense of dependence that is established: the individual wants to be in a certain context or achieve a performative level, but cannot reach it except by resorting to the device. Addiction is not usually viewed as deriving from the fact that technology has opened up new dimensions of experience: it is easier, albeit erroneous, to consider the body as a deficient entity that demands compensation through the various technological instruments. This is a fallacy, which leads us to consider addiction as a condition *ab origine* when, on the contrary, it is the outcome of a somatic adaptation mechanism that originates in techno-poiesis itself.

Overcoming the notion of somatic incompleteness requires a shift from perceiving technological implementation as a response to deficiency to embracing a different perspective that acknowledges the process of complementation with the device, resulting from the experience of technomediation. This paradigm shift is essential for understanding the profound ontological implications of technology and cultivating a new culture of *techne*. Dependency does not stem from the device filling a perceived gap, but rather from the body's innate inclination to complement and integrate with the device (Cerqui, 2002). In other words, every technical apparatus fosters hybridisation, not only by enabling hybrid performances, but also by inducing changes and configurations within the body to establish a complementary relationship with the device. The technical apparatus is hybridising because it intertwines itself with the body, while simultaneously making the organism receptive to its integration. As mentioned, it is easier for the body to adapt to technology than the other way around, both because organisms have greater plasticity and because the resulting hybrid function is never fully predictable. This calls for a profound transformation in our understanding not only of human nature, but also of the dynamics of technomediation.

Indeed, the experience of technology does not lead to a detachment or emancipation of the body from the world; instead, it brings about a *metamorphosis of the flesh*, resulting in a different relationship and new connections with the external environment (Tosun and Lajunen, 2009). The device always has a viral-like impact – if we were to liken the body to a cell – and this infection-like effect creates a dependency, similar to how eukaryotic cells cannot function without their mitochondria. In this transformation of the subject's identity interface, we cannot overlook the profound connection established with the technical-technological referent. It forms a prospective and performative bond that reinforces the human being's needs. The body, in its complementary state, already anticipates the assistance of the device, eventually leading to a condition of heteronomy where the body's structure implies an external contribution (Kuss and Griffiths, 2012). Therefore, technomediation is not external to the body but rather inherent to it, much like how the gills of a fish anticipate the presence of dissolved oxygen in water.

While we perceive this dependence, we often misinterpret it. This misunderstanding gives rise to an explanatory distortion, leading us to view technology as a compensatory technocultural support. In other words, we mistake a consequence (an *ex-post* condition) for a cause (an *a-priori* need). This blunder can be attributed to a common explanatory bias. Complementation, in essence, is not a form of compensation, which implies the correction of a deficiency. Instead, it refers to the body's inclination to achieve higher levels of performance by adapting itself to performative hybridisation. There is a genuine dependence on functional conformation, wherein the body has already positioned itself at a different performative level compared to before. The introduction of any cultural or technical apparatus has the power to shape human perspectives and performance, significantly impacting the adaptive conformation of the body in relation to the device (Rivoltella and Rossi, 2019). The act of hybridisation is, in fact, an inclusive process, as discussed earlier in relation to somatisation. Individuals assimilate the hybrid qualities achieved through an apparatus that is now an integral part of them. Whenever people invent a technological tool, it becomes integrated into their body, resulting in the somatisation of its properties. Consequently, the somatic configuration of the individual evolves with the aim of complementing the tool. We could even describe this process as one of hospitality, wherein the realisation of the hybrid function can only occur if the graft takes hold and is not rejected. In other words, the somatic dimension must accept, integrate, and organically conform to the device.

Nevertheless, the body is frequently subjected to significant stress or coercion from devices, leading to the development of addictions resulting from the organism's inability to effectively incorporate, cushion, integrate, and systematise the technical apparatus. These new forms of addiction arise from conditions of discomfort that intensify an individual's exposure to techno-perturbation. Notably, we observe the emergence of various addictions linked to the development of technologies that surpass human control. A clear example is that of barcodes, which are designed to be read by other technologies rather than human eyes (Floridi, 2014). The mere lack of understanding regarding how a particular machine operates, with access limited only to its performance, erodes human control and fosters dependence. Even in the realm of financial trading, speed has reached such levels that only networked algorithms can keep up. Addiction also arises from the imperative of constant connection, where individuals find themselves unable to disconnect without bearing significant consequences (Leonhard, 2016). If, for instance, we choose to turn off our smartphone and an unfortunate accident occurs to a loved one, we might feel accountable. Moreover, we have lost the privilege of ignoring emails, WhatsApp messages, shared information, or comments on posts. An immediate response is expected, operating under the assumption that everyone is always connected and available to address requests or read messages. This hyper-connectedness also amplifies dependence on the judgement of others, prioritising the social self over the private dimension and resulting in constant exposure.

Technostress refers precisely to the discomfort or chronic fatigue that accompanies excessive connectivity (Brod, 1982). Common symptoms include restlessness, fluctuating arousal, anxiety, difficulty falling asleep, frequent awakenings during the night, and decreased libido. Indeed, it is well-known that stress triggers physiological mechanisms involving the production of cortisol and aldosterone, which decrease the release of sex hormones such as testosterone and oestrogen. This hormonal imbalance can have detrimental effects on muscle and bone metabolism, as well as on the sex drive. Prolonged exposure to this stressful condition (distress) can significantly reduce immune response, making individuals more susceptible to infections by microorganisms and contributing to the development of psychosomatic conditions. Hyperconnectivity intensifies the release of attentional neuromodulators like glutamate and acetylcholine, leading to a state of overexcitation that hampers mental rest. This process has been found to have neurotoxic effects, particularly on the hippocampus, impairing memory mechanisms (Carlson, 2013). Additionally, there is an excessive activation of alert systems, particularly the amygdala, which results in heightened emotional sensitivity, including irritability and anxiety. The impact on the hypothalamus increases the production of vasopressin, a hormone associated with aggression.

One of the consequences of hyper-connectedness is a new type of separation anxiety, or nomophobia, characterised by a fear of being excluded or missing out on something. In severe cases, nomophobia can lead to panic-like symptoms such as dizziness and shortness of breath (D'Agata, 2008). The disconnection syndrome primarily affects vulnerable individuals, including those with low self-esteem, people living in poverty or precarious conditions, individuals with social difficulties, or those with heightened sensitivities. It manifests through the emergence of obsessive thoughts, compulsive behaviours, and symptoms commonly associated with anxiety disorders. The incessant need to be constantly connected intensifies the individual's exposure to the online world, fostering a desire to feel accepted, validated, and gratified. Also, the pursuit of public presence in the digital realm often results in meticulous image management, reflecting a state of discomfort. There is a strong desire to constantly monitor how others perceive us, which can create a self-imposed prison for individuals.

Furthermore, hyperconnectivity can lead to a decline in social skills, sometimes reaching pathological levels as seen in the phenomenon of Hikikomori in Japan, where young people, primarily teenagers, isolate themselves at home for extended periods of time (Fogel et al., 2007). While not all individuals reach such extreme conditions, it is evident that in recent decades, challenges have emerged that can be attributed to a significant decline in social interaction. These difficulties include social anxiety in face-to-face relationships, the inability to effectively communicate in person, a lack of relational competence, and difficulties in forming meaningful friendships. Therefore, we can identify new forms of social isolation stemming from hyperconnectivity, which contribute to detachment from the real world and peer relationships.

Some common forms of addiction arising from hyperconnectivity include: i) Vamping: This refers to the habit of staying awake late into the night to engage in activities such as chatting, posting, checking social networks, replying, or playing online games. ii) Like addiction: This entails a constant need to receive likes and shares, leading to compulsive behaviour such as continuously posting messages, photos, and videos to seek appreciation. In both cases, the influence of the dopaminergic system, which undergoes reconfiguration through online interactions, is evident (Rosen, 2012). The user's interest, expectations, exploratory tendencies, and seeking mechanisms are all influenced by the gratification processes, particularly the element of randomness in rewards and the search for surprises. The state of anticipation and uncertainty regarding outcomes creates a neurophysiological similarity, in terms of its impact, between the user's interaction with online platforms and other forms of addiction, such as gambling. Internet Addiction Disorder (IAD), a term coined by psychiatrist Ivan Goldberg in 1995, describes compulsive behaviour characterised by excessive and persistent Internet use, resulting in emotional and cognitive disorders (Skues et al., 2012). This addiction can manifest in various ways, including cyber relational addiction or cybersex addiction. Individuals may seek identity affirmation through accumulating a large number of friends and followers, as well as a constant need to stay informed, which can lead to information overload or a compulsive wish for news updates.

Digital technologies, on the other hand, foster a climate of networked participation that extends beyond the boundaries initially set by the individual. They serve as multipliers of interactions that transcend the realm of the interpersonal, generating viral content that can replicate endlessly without limitations or constraints (Zhao et al., 2008). While online presence offers opportunities for relationships, it extends far beyond the realm of personal interactions. Any content shared in the digital sphere takes on a viral nature, functioning as a replicator that spreads based on user access. Consequently, content on the Internet becomes a shared resource, accessible to all, and cannot be confined within the boundaries of privacy (Trepte and Reinecke, 2013). As a result, each connection reinforces itself and fuels the individual's need to be increasingly present on digital platforms. However, managing multiple profiles, often with conflicting or non-overlapping identities, depletes self-narration resources and tips the scales towards seeking external validation and participating on multiple fronts simultaneously. It's no wonder, then, that the cohesive internal identity of the individual is easily lost in the process (Raskin and Terry, 1988).

From these considerations, it can be inferred that hyper-connection leads to the depletion of the subject's cognitive resources. The constant bombardment of attentional demands results in decreased reflective responses and a tendency towards impulsivity, generality, and distraction. Prolonged exposure to the online world creates conditions of stress, frustration, overstimulation, and irritability, often leading to detachment from reality or even hallucination. These factors contribute to the fragmentation of identity as they diminish the subject's capacity for cohesion and integration. Epidemiological research conducted in recent decades reveals a significant decline in cognitive qualities such as memory and reflexivity, which play a crucial role in maintaining a continuous and coherent identity. These qualities, even in the context of growth and change, serve as the best defence against fragmentation. It is also important to note the role of sleep in identity organisation. Hyper-connectedness, among other factors, contributes to forms of insomnia and frequent sleep disruptions (Wood and Scott, 2016). Micro-awakenings and interrupted sleep during the night result in a constant feeling of fatigue and reduced attention span during waking hours. The use of electronic devices, especially in the evening, has been identified as one of the leading causes of disrupted deep sleep and sleep interruption. However, few realise that these disturbances not only lead to continuous tiredness and decreased cognitive capacity but also have an impact on memory and disrupt the sense of a cohesive self.

Social platforms amplify the individual's need for social validation and foster a tendency to segment one's audience based on the feedback received. This process inevitably creates difficulties in genuine dialogue and fuels a narcissistic inclination. Consequently, we encounter the paradox of a dispersed identity that is nevertheless egocentric in its relationships. Rather than unifying the fragments of the self through a connective framework, hyper-connectedness encourages their dispersion. The vast array of potentially accessible spaces presented to individuals stimulates their desire to participate, leading to compulsive behaviours such as creating new profiles and constantly sharing pictures, videos, and messages. The allure of immediate and superficial content makes it highly shareable, contributing to the success of platforms like Twitter and Instagram. Similar to advertising mechanisms that exploit desires and the wish to conform, fostering consumerist compulsions, the infosphere represents a new frontier of possibilities that never fully satisfy the individual, driving them to be constantly present and engaged. The narcissistic mechanism creates a narcotic prison, ensnaring individuals within solipsistic cells as they limit themselves to sending out messages in bottles, desperately seeking attention and affirmation (Buffardi and Campbell, 2008).

All the aforementioned sources of discomfort undeniably erode the well-being of individuals, who increasingly seek compensation, paradoxically turning to the very devices that caused these issues. It is not coincidental that those who are most vulnerable and lack other coping mechanisms are the ones most susceptible to these new forms of addiction. The greater the individual's state of imbalance, the stronger the addictive effect becomes. Another concerning aspect is the exposure of young people to these digital media, a phenomenon that has raised alarm, especially due to the accelerated impact resulting from the COVID-19 pandemic. Many young people have experienced lockdown during crucial stages of social development, such as pre-adolescence and puberty. For two years, they relied on occasional, fragmented, and predominantly online interactions with their peer group. This situation has further emphasised the already significant influence of the digital world on their lives while causing a steep decline in face-to-face social connections. Indeed, it is important to highlight that while the pandemic has caused a profound disruption in the lives of adults, it has had an even more substantial impact on young people, as two years represent a significantly larger portion of their lives.

Predicting the long-term repercussions of this event is challenging. However, it is clear that the excessive immersion in digital experiences among younger people is a problem that can no longer be ignored, as its consequences are already becoming evident. Adopting a critical mindset and maintaining a balanced approach towards the opportunities presented by technology does not imply endorsing technophobic or neo-Luddite attitudes. It simply means recognising that every transformation, including technological advancements, carries both opportunities and risks. Therefore, practising moderation remains the most effective antidote in navigating this complex landscape.

CONCLUSION

To view the relationship between the body and technology as a process of hybridisation, a metamorphic relationship that involves a reciprocal attribution of functional value and gives rise to new characteristics, we must move away from a purely instrumental understanding of technology. In an instrumental view, the device remains external, inert and passive, under the complete control of the human being who determines its function and decides if, when, and how to use it. However, this view has never truly reflected reality. Until the 20th century, one could still cling to the humanist illusion of an impermeable body utilising instruments to achieve predetermined ends unaffected by the presence of the device. Today, this belief is unequivocally contradicted by the advancements in technology. Adhering to such an interpretative paradigm hinders our ability to comprehend the diverse scenarios, encompassing both risks and opportunities, that the development of the technosphere as a whole places before us.

Another incorrect way to interpret technology is to personify it, portraying it as either an evil entity seeking to destroy humanity or as a benevolent force that will save us from all harm. Firstly, it is important to recognise that there is no such thing as a singular "technology," but rather a *technospheric dimension* within which a new manifestation of the human being emerges. This dimension can be seen as a niche as it is understood in ecology: therefore, an ecological approach ought to be used in relation to technology as well. A species is not simply an inhabitant of an environment; it is a unique perspective on reality, a distinct way of being that arises from the symbiotic relationship between the subject and the world. The niche is the result of this relationship, where both elements mutually influence and modify each other. Thus, we need to move away from an essentialist view of the human being and expand our understanding of the individual by transcending anthropocentric boundaries and embracing a broader hybrid-relational space.

Taking a technophysiological perspective means placing hybridisation processes at the forefront in order to understand the ontopoietic shifts that have made the human being an inherently meta-stable entity. This approach goes beyond merely highlighting bodily changes in relation to various technological mediums and goes beyond a descriptive analysis with a clinical, pathogenetic, or epidemiological focus. *Technophysiology aims to provide insights into a new somato-technical phenomenology that reveals the hybrid nature of the human experience.* It seeks to overcome the solipsistic view that has characterised humanist anthropocentrism, a paradigm underlying the major crises of our time. By clinging to the notion of human purity, in fact, this view has embraced the myth of biological incompleteness and the theory of Promethean exoneration, relying on compensatory interpretations while neglecting the profound implications of the relationship between the body and technology. Today, with the emergence of Artificial Intelligence, digital technologies, the Metaverse, biosynthesis, and quantum devices, such a narrow perspective becomes dangerously limiting.

Beyond the inherent risks that accompany every technology, the true peril lies in our failure to comprehend the far-reaching consequences of technomediation. This includes the countless transformations brought about by hybridisation: direct changes through the impact on the body, as well as indirect modifications in the nature of external influences. In other words, the qualities manifested by humans cannot be deduced through internal introspection, as the lingering humanist approach still attempts to do. Instead, these qualities emerge as a result of the relationship with the technospheric niche. However, this emergent nature extends beyond the mere characterisation of human expressions, as it influences phenotypic variations, epigenetic inheritance, phylogenetic trajectories, and the ontogenetic development of individuals. Technophysiology goes beyond merely examining pathological alterations caused by interaction with devices, demonstrating that the body cannot be conceived as an entity that realises and regulates itself based solely on genetic essence.

Human beings emerge as a product of the integration of their apparatuses, for we do not have a body but *are* a body. This integration occurs through a continuous interaction with the surrounding world, mediated by our niche, in a fractal process that can be observed at various stages of organismic development. Even before becoming an embryo and eventually an individual, the zygote already carries within itself the imprints of its niche. I refer to this as *heteronomy*, an inherent morphopoietic condition that anticipates the shaping influence of external factors. Thus, hybridisation is an ontological prerequisite. However, it is important to note that this condition is not a mere juxtaposition, but rather an expression. The body represents a rough draft, susceptible to all forms of inheritance, including those coming from the niche, while also remaining open to metamorphosis precisely because of its inherent variability. Therefore, it is crucial to avoid considering the human being as a static entity detached from fluctuating variations. We must reject the presumption of humanism, which seeks to maintain a fixed identity by stripping away any external elements, as depicted in the Vitruvian Man.

Technophysiology reveals that every technical apparatus is embodied because, even before inducing changes in the body, it is tasked with translating variability into an overall form. In contrast to the humanist perspective where internal reception remains limited and supplemental, separate from the organism's dynamics and treated as an external entity solely providing a given performance, a new techno-physiological conception erases this disjunction. It views the body as a dynamic co-factoring expression capable of constructing hybrid processes, not due to any deficiency but precisely because of its redundancy. Redundancy thus serves as the fundamental principle underlying the virtuality of the body and, more broadly, of the organic realm, manifesting as variability. This implies that hybridisation is already inherent, constantly evolving and open to possibilities, so that technopoiesis does not depart or diverge from the ecological mechanisms of living organisms. We are not confined to a static reality; nature is in a perpetual state of transformation, with moments of apparent stability interspersed with sudden accelerations. Life itself can be understood as the translation of radiant energy into diachronic accumulation systems. Thus, humanity must rediscover its connection to the biological convivium, also through its relationship with technology.

What is emerging is a *new perspective on humanity* that transcends the narrow focus on the human being as a solipsistic entity and instead recognises the hybrid nature of our existence, with an awareness of our eco-ontological status. It is not merely a call for humility, reminding us of our profound dependence on the world and challenging our arrogant claims of self-determination, but also an opportunity to examine the consequences of technological progress with greater clarity. We should neither fear technology nor deceive ourselves into believing that we have complete control over it, dismissing its issues with a simple phrase like "it depends on how you use it." Reality is far more complex than that, because each innovation not only affects our operational capacities but also transforms our very being. Understanding this does not require relinquishing our capacity for decision-making; instead, it invites us to deepen our awareness and enhance our ability to interpret phenomena. *Anthropocentrism is not a privileged position* that humanity must renounce out of moral obligation. It is a state of ignorance that must be overcome – a blindfold that needs to be lifted.

Keeping this blindfold on, while developing technologies with self-programming capabilities and intricate control mechanisms over social and individual orientations, is a risky proposition to say the least. The problem lies in the persistent bias that permeates our thinking about technology: the notion that we have complete control over its use. It is challenging to let go of the idyllic view of tools as mere enhancers of our capabilities. Within this perspective, the human remains the constant, the fixed point, the axis around which everything revolves. Even in transhumanist discourse, one can observe this ambiguity. Absolute values are postulated for the envisioned post-humanity, neglecting the fact that such values are contingent upon considering the human as a stable entity. This gives rise to inherent contradictions, such as attempting to identify an absolute while diluting the very foundation that should support it – a paradox I have termed "anthropocentrism without *anthropos*." Therefore, transcending the framework of humanism by embracing a posthumanist transition becomes an essential prerequisite for navigating the accelerated processes of technomediation. Although we have always been cyborgs, until recently we could still ignore or feign ignorance about this fact, comforting ourselves with the anthropocentric illusion that everything revolved around us, and that human ingenuity crafted tools to increase our power and eventually emancipate us from nature. Today, such thinking is no longer simply misleading. It poses a serious threat to our future.

BIBLIOGRAPHY

Abouheif, Ehab, et al. 2014. "Eco-evo-devo. The time has come". Adv. Exp. Med. Biol., 781: 107-125.

Abram, John. 2007. The Language of Winnicott. London: Routledge.

Adams, Fred, and Aizawa, Ken. 2001. "The bounds of cognition". Philosophical Psycology, 14, 1: 43-64.

Agamben, Giorgio. 2004. The open. Man and Animals. Stanford: Stanford University Press.

Ainsworth, Mary, et al. 1978. Patterns of Attachment. Hillsdale: Erlbaum.

Alborghetti, Roberto. 2022. De(generazione) digitale. Roma: Funtasy Editrice.

Alfano Miglietti, Francesca. 2008. Identità mutanti, dalla piega alla piaga. Milano: Mondadori.

Allenby, Braden R., and Sarewitz, Daniel. 2011. The Techno-Human Condition. Cambridge: The MIT Press.

Alter, Adam. 2018. Irresistible. New York: Penguin Books.

Amoretti, Maria Cristina and Manzotti, Riccardo. 2012. "Esternalismi". Rivista di Filosofia, 1: 41-68.

Amzallag, Nissim. 2020. "Technopoiesis. The forgotten dimension of early technique development". Philosophy & Technology, 34.

Anders, Günther. 1980. Die Antiquiertheit des Menschen, Voll I. München: C.H. Beck.

Anderson, Craig, et al. 2003. "The influence of media. Violence on youth". Psychological Science in the Public Interest, 4: 81-110.

Anderson, Janna Q., and Rainie, Lee. 2012. Millennials will benefit and suffer due to their hyperconnected lives. https://www.pewresearch.org/internet/2012/02/29/millennials-will-benefit-and-suffer-due-to-their-hyperconnected-lives/

Appelbaum, L. Gregory, et al. 2013. "Action videogame playing is associated with improved visual sensitivity, but not alterations in visual sensory memory". Attention, Perception, & Psychophysics, 75, 6: 1161-1167.

Armstrong, David M. 1981. What is consciousness? In The Nature of Mind, edited by John Heil. New York: Cornell University Press.

Arnett, Jeffrey J. 2005.Emerging adults in America: coming of age in the 21st century. Washington DC: American Psychological Association Press.

Art, Jason. 2022. Bias cognitivi. Italy: Amazon Print.

Arthur, Brian W. 2009. The Nature of Technology. New York: Free Press.

Aschoff, Jürgen. 1979. "Circadian Rhythms: Influences of Internal and External Factors on the Period Meausured in Constant Conditions". Zeitschrift für Tierpsychologie, 49, 3: 225-249.

Atkinson Taylor J., and Halfon, Marc S. 2014. "Regulation of Gene Expression in the Genomic Context". Computational and Structural Biotechnology Journal, January 29.

Attwood, Terri, et al. 2011. Concepts. Historical Milestones and the Central Place of Bioinformatics in Modern Biology: A European Perspective. Bioinformatics. In Trend and Methodologies, edited by Mahmood A. Mahdavi, 3-38. Shanghai: InTech.

Aud Sissel, Hoel. 2018. "Operative Images. Inroads to a New Paradigm of Media Theory". In Image-Action-Space: Situating the Screen in Visual Practice, edited by Luisa Feiersinger, Kathrin Friedrich and Moritz Queisner, 11-28. Berlin, Boston: De Gruyter.

Bachimont, Bruno. 2010. Le sens de la technique. Encre Marine.

Bailenson, Jeremy. 2018. Experience on Demand. New York: W.W. Norton & Company.

Baker, David J.P. 1994. Mothers, Babies, and Disease in Later Life. London: BMJ Publishing Group.

Baker, Lynne R. 2011. Persons and Bodies. A Constitution View. Cambridge: Cambridge University Press.

Balibar, Étienne. 2018. "Philosophies of the Transindividual: Spinoza, Marx, Freud". Australian Philosophical Review, 2: 5-25.

Bamford, Sim. 2012. "A framework for approaches to transfer of a mind's substrate". International Journal of Machine Consciousness, 4, 1: 23–34.

Barbrook, Richard. 2007. Imaginary Futures. London: Pluto Press.

Barker, G. et al. 1994. "Placental water content and distribution". International Journal of Gynecology & Obstetrics, 47, 2: 193-194.

Barnstone, Tony. 2000. "Technology as Addiction". Technology and Culture, 41, 1: 190-193.

Barresi, Michael J., and Gilbert, Scott F. 2020. Development Biology. Oxford: Oxford University Press.

Barry, Patricia D. 2002. Mental Health & Mental Hill. New York: Lippincott.

Bartlett, Frederic C. 1932. Remembering. Cambridge: Cambridge University Press.

Bartness, T.J., et al. 1993. "The timed infusion paradigm for melatonin delivery: What has it taught us about the melatonin signal, its reception,

and the photoperiodic control of seasonal responses?". Journal of Pineal Research, 15: 161-190.
Bateson, Gregory. 1976. Steps to an Ecology of Minds. New York: Ballantine Books.
Bateson, Gregory. 2005. "The return of the whole organism". J. Biosci., 30: 31-39.
Bateson, Patrick, and Gluckman, Peter. 2011. Plasticity, Robustness, Development, and Evolution. Cambridge: Cambridge University Press.
Bateson, Paul P. G. and Klopfer, Peter H. 1973. Perspective in Ethology. Dordrecht: Kluwer Academic/Plenum Publishers.
Bauman, Zygmunt. 1998. Globalization: The Human Consequences. New York: Columbia University Press.
Bechtel, William et al. 1998. The Life of Cognitive Science. In A Companion to Cognitive Science, edited by William Bechtel and George Graham. Oxford: Basil Blackwell.
Beck, Benjamin B. 1980. Animal Tool Behavior. New York: Garland Publishing.
Belamghari, Mohamed. 2020. "The Fragmentation of Identity in the Age of Glocalization". Sage Open, 10: 1-12 https://www.researchgate.net/publication/342332455_The_Fragmentation_of_Identity_Formation_in_the_Age_of_Glocalization.
Benasayag, Miguel, and Schmit, Gérard. 2003. Les passions tristes. Paris: Éditions La Découverte.
Benasayag, Miguel. 2015. El cerebro aumentado, el hombre disminuido. Buenos Aires: Paidós.
Bergson, Henri. 1907. L'évolution Créatrice. Paris: Presses Universitaires France.
Berridge, Kent C. 2007. "The debate over dopamine's role in reward, The case for incentive salience". Psychopharmacology, 191: 391-431.
Berridge, Kent. C., and Kringelbach, Morten L. 2015. "Pleasure in the Brain". Neuron, 86, 3: 646-664.
Biuso, Alberto G. 2020. Animalia. Catania: Villaggio Maori.
Blackburn, Simon. 2014. Mirror, Mirror: The Uses and Abuses of Self-Love. Princeton, NJ: Princeton University Press.
Block, Jerald J. 2008. "Issues for DSM-V: Internet Addiction". The American Journal of Psychiatry. 165, 3: 306-307.
Bocchi, Gianluca, and Ceruti, Mauro. 2007. La sfida della complessità. Milano: Bruno Mondadori.
Bodei, Remo. 2016. Limite. Bologna: Il Mulino.
Bojanova, Irena. 2014. "The Digital Revolution: What's on the Horizon?". IT Professional, 16, 1: 8-12.

Bolk, Louis. 1926. Das Problem der Menschwerdung. Jena: Fischer.
Bonderud, Doug. 2021. "Allostasis: Balancing the Books of Brain-Body Budgets". Now. Powered by Northrop Grumman, 2021-05-05.
Borgmann, Albert. 1984. Technology and the Character of Contemporary Life. Chicago: The University of Chicago Press.
Borin, Paolo. 2005. La mano e la mente. Roma: Carocci.
Bostrom, Nick. 2018. Superintelligenza. Tendenze, pericoli, strategie. Torino: Bollati Boringhieri.
Bowlby, John. 2015. The Making and Breaking of Affectional Bonds. New York: Routledge.
Bradshaw, A. D. 1973. "Environment and phenotypic plasticity". Brookhaven Symposium in Quantitative Biology, 25: 74-94.
Braidotti, Rosi. 2013. The Posthuman. Cambridge: Polity.
Brambell, Roger. 1965. Report of the Technical Committee to Enquire into the Welfare of Animals kept under Intensive Livestock Husbandry Systems. Her Majesty's Stationary Office. London.
Bramble, Dennis M., and Daniel E., Lieberman. 2004. "Endurance running and the evolution of Homo". Nature, 432: 345-352.
Breggin, Peter R., and Breggin, Ginger Ross. 2014. Talking Back to Prozac: What Doctors Won't Tell You Prozac and the Newer Antidepressants. New York: Open Road Media.
Brey, Philip. 2000. "Recent Developments in Dutch and European Philosophy and Ethics of Technology". Techné: Research in Philosophy and Technology, 24, 4: 36-42.
Brinster, Ralph L., and Palmiter, Richard D. 1982. "Dramatic growth of mice that develop from eggs microinjected with metallothionein-growth hormone fusion genes". Nature, 300: 611-615.
Broadhurst Susan, and Machon, Josephine. 2012. Identity, Performance and Technology. London: Palgrave Macmillan.
Brod, Craig. 1982. "Managing Technostress: Optimizing the Use of Computer Technology". Personal Journal, 61, 10: 753-757.
Broom, Donald M., and Johnson, Ken G. 1993. Stress and Animal Welfare. Netherlands: Springer.
Brosnan, M. J., 2002. "The Impact of computer anxiety and self-efficacy upon performance". Journal of Computer Assisted Learning, 14, 3: 223-234.
Bruner, Jerome. 1986. Actual Minds, Possible Worlds. Cambridge: The Jerusalem-Harvard Lectures. Harvard University Press.
Buck, Linda B. 1996. "Information coding in the vertebrate olfactory system". Annu Rev Neurosci, 19: 517-544.

Buffardi, Laura E., and Campbell, W. Keith. 2008. "Narcissism and social networking Web sites". Personality and Social Psychology Bulletin, 34: 1303-1314.

Burge, Tyler. 1979. "Individualism and the Mental". Midwest Studies in Philosophy, 4: 73-121.

Burke, M., et al. 2010. "Social network activity and social well-being". Proceedings of the SIGCHI Conference on Human Factors in Computing System, 3: 203-220.

Buss, David M. edited by. 2005. The handbook of evolutionary psychology. Hoboken: John Wiley & Sons.

Bussagli, Marco. 1999. Rome. Art & Architecture. New York: Konemann.

Byung-Chul, Han. 2018. The Expulsion of the Other: Society, Perception and Communication Today. Cambridge: Polity.

Campbell, Donald T. 1974. Evolutionary Epistemology. Chicago: Open Court Publishing Co.

Campmans-Kuijepers, Mario J.E., and Dijkstra, Gerard. 2021. "Food and Food Groups in Inflammatory Bowel Disease (IBD): The Design of the Groningen Anti-Inflammatory Diet (GrAID)". Nutrients, 13, 4: 1067.

Capucci, Pier Luigi. 1994. Il corpo tecnologico. Bologna: Baskerville.

Carlson, Neil R. 2013. Physiology of behavior. Essex: Pearson Education.

Caronia, Antonio. 1985. Il cyborg. Saggio sull'uomo artificiale. Roma-Napoli: Edizioni Teoria.

Carr, Nicholas. 2010. The Shallows. What the Internet is Doing to Our Brains. New York: W.W. Norton.

Carroll, Sean B. 2005. Endless Forms Most Beautiful. New York: Baror International.

Castells, Manuel. 1999. "Information Technology, Globalization and Social Development". UNRISD, Discussion Paper 114, September.

Castells, Manuel. 2007. "An introduction to the information age". Analysis of Urban Change, Theory, Action, 2, 7: 6-16.

Castells, Manuel. 2009. The Rise of the Network Society. Oxford: Blackwell Publishing.

Cerqui, Daniela. 2002. "The future of humankind in the era of human and computer hybridization: An anthropological analysis". Ethics and information Technology, 4: 101-108.

Chaitin, Gregory. 2010. Leibniz, Complexity and Incompleteness. In Causality, Meaningful Complexity and Embodied Cognition, edited by Arturo Carsetti, 127-133. Cham: Springer.

Changeux, Jean-Pierre. 1986. Neuronal Man. New York: Oxford University Press.

Changizi, Mark A., et al. 2008. "Perceiving the Present and a Systematization of Illusions". Cognitive Science, 32: 459-503.
Cheever, Nancy, et al. 2014. "Out of sight is not out of mind: The impact of restricting wireless mobile device use on anxiety levels among low, moderate and high users". Computers in Human Behavior, 37: 290-297.
Chirchir, Habiba, et al. 2015. "Recent Origin of Low Trabecular Bone Density in Modern Humans". PNAS, 112: 366-371.
Chislenko, Alexander. 1995. http://www.lucifer.com/~sasha/home.html.
Clark, Andy and David, Chalmers. 1998. "The extended mind". Analysis, 58, 1: 7-19.
Clark, Andy. 1989. Microcognition: Philosophy, Cognitive Science and Parallel Distributed Processing. Cambridge: The MIT Press.
Clark, Andy. 2003. Natural-Born Cyborg: Minds, Technologies, and the Future of Human Intelligence. Oxford: Oxford University Press.
Clarke, Roger. 1994. "The Digital Persona and its Application to Data Surveillance". The Information Society, 10, 2: 77-92.
Clasen, Mathias et al. 2019. Adrenaline junkies and white-knucklers: A quantitative study of fear management in haunted house visitors. "Poetics", 73: 61-71.
Clifford, Brian. 2018. A brief history of woodturning. "The Woodturner's Workshop". Woodturners' Guild of Ontario.
Clynes, Manfred E., and Kline, Nathan S. 1960. "Cyborg and Space". Astronautics, 26, 7: 74-76.
Colegrave, Nick, and Collins, Sinead. 2008. Experimental evolution: experimental evolution and evolvability. "Heredity", 100, 5: 464-470.
Colvile, Tom. 2016. The Great Acceleration: How the World is Getting Faster, Faster. Londra: Bloomsbury Publishing.
Cornoldi, Cesare. 1997. Metacognizione e Apprendimento. Bologna: Il Mulino.
Corruccini, Robert S., and Beecher, R.M. 1984. "Occlusofacial morphological integration lowered in baboons raised on soft diet". J Cran Genet Dev Bio, 4: 135-142.
Costa, Mario. 1990. Il sublime tecnologico. Salerno: Edisud.
Crain David A., and Guillette, Louis J. 1998. "Reptiles as models of contaminant-induced endocrine disruption". Animal Reproduction Science, 53, 1-4: 77-86.
Cregan-Reid, Vybarr. 2018. Primate Change. How the world we made is remaking us. London: Cassell.
Cross, Gary. 2008. Men to Boys: The Making of Modern Immaturity. New York: Columbia University Press.

Cumming, Hope M., and Vandewater, Elizabeth A. 2007. "Relation of Adolescent videogame play to time spent on other activities". Archives of Pediatrics & Adolescent Medicine, 161, 7: 684.

D'Agata, Charlie. 2008. "Nomophobia: Fear of being without your cell phone". CBS News, april 3.

D'Alessandro, Paolo, and Potestio, Andrea, edited by. 2006. Filosofia della tecnica. Milano: LED Edizioni Universitarie.

Dainoff, Mavin J. editors. 2007. Ergonomics and Health Aspects of Work with Computers. Heidelberg: Springer.

Damasio, Antonio et al., 1996. "The somatic marker hypothesis and the possible functions of the prefrontal cortex". Philosophical Transactions of the Royal Society of London. Series Biological Sciences, 351, 1346: 1413-1420.

Damasio, Antonio. 1994. Descartes' Error: Emotion, Reason, and the Human Brain. New York: HarperCollins.

Davidson, Theresa, and Farquhar, Lee. 2020. "Prejudice and Social Media: Attitudes Toward Illegal Immigrants, Refugees, and Transgender People". In Gender, Sexuality and Race in the Digital Age, edited by Compton, D'Lane R. and Andrea P. Herrera, D. Nicole Farris, 151-167. Cham: Springer.

Davies, Mark J. 2022. Metaverse. Chicago: Independently Publisher.

Davis, Henry Gassett. 1867. Conservative Surgery. New York: D. Appleton & Co.

Dawkins, Richard. 1976. The Selfish Gene. Oxford: Oxford University Press.

De Clercq, Erik and Kern, Earl R. 2003. Handbook of Viral Bioterrorism and Biodefense. New York: Elsevier.

De Kerckhove, Derrick. 1991. Brainframes: Technology, Mind and Business. Utrecht: Bosch & Keuning.

De Kerkhove, Derrick. 1998. Connected Intelligence: The Arrival of the Web Society. London: Kogan Page.

Debes, Remy. 2017. Dignity. Oxford: Oxford University Press.

Deci Edward L., and Flaste, Richard. 1996. Why We Do What We Do: Understanding Self-Motivation. New York: Penguin.

Deci, Edward L., and Ryan, Richard M. 1985. "Conceptualization of Intrinsic Motivation and Self-Determination". In Intrinsic Motivation and Self-Determination in Human Behavior, edited by Edward L. Deci and Richard M. Ryan, 11-40. Cham: Spring.

Dehaene, Stanislas. 2020. How We Learn. New York: Viking.

Deleuze, Gilles. 1994. Difference and repetition. London: Athlone Press.

Deleuze, Gilles. 2010. Cosa può un corpo? Lezioni su Spinoza. Verona: Ombre Corte.
Demichelis, Lelio. 2022. Marx, Alienation and Techno-Capitalism. Cham: Springer International Publishing.
Deng, Li, and Yu, Dong. 2014. Deep Learning: Methods and Applications. Delft: Now Publisher.
Dennett, Daniel C. 1996. Kinds of minds. Toward an Understanding of Consciousness. London: Weidenfeld and Nicolson.
Derrida, Jacques. 2008. The animals that therefore I am. New York: Fordham University Press.
Descola, Philippe. 2005. Par-delà nature et culture. Paris: Gallimard.
Deutsch, David. 1997. The Fabric of Reality. London: Penguin Books.
Di Francesco, Michele, and Marraffa, Massimo. 2009. Il soggetto. Scienze della mente e natura dell'io. Milano: Mondadori.
Diamanti-Kandarakis, Evanthia, and Gore, Andrea C., edited by. 2011. Endocrine Disruptors and Puberty. Totowa: Humana Press.
Dickson. E. J. 2020. "New Research Explains Why Boomers Share So Much Fake News". Rolling Stone, 27 May.
Dormehl, Luke. 2014. The Formula. Los Angeles: TarcherPerigree.
Dossey, Larry. 2014. "FOMO, Digital Dementia, and Our Dangerous Experiment". Explore, 10, 2: 69-73
Douglas, Angela E. 2014. The Symbiotic Habit. Princeton: Princeton University Press.
Douglas, Ian. 2020. Is Technology Making Us Sick? London: Thames & Hudson.
Drake, Christopher, et al. 2004. Vulnerability to stress-related sleep disturbance and hyperarousal. Sleep, 27, 2: 285-291.
Dreyfus, Hubert L. 1992. What Computers Still Can't Do: A Critique of Artificial Reason. Cambridge: The MIT Press.
Dupré, John. 2002. Human Nature and the Limits of Science. Oxford: Oxford University Press.
Dyson, George B. 1997. Darwin Among the Machines. New York: Perseus Books.
Eco, Umberto. 1965. Apocalittici e integrati. Milano: Bompiani.
Edelman, Gerald M. 1972. "Antibody Structure and Molecular Immunology". Nobel Lecture.
Edelman, Gerald M. 1987. Neural Darwinism. New York: Basic Books.
Eggebrecht Adam T. et al. 2014. "Mapping distributed brain function and networks with diffuse optical tomography". Nature Photonics, 8: 448-454.

Eibl-Eibesfeldt, Irenäus. 1989. Human Ethology. New York: Aldine De Gruyter.
Eriksen, Thomas Hylland. 2001. Tyranny of the Moment. London: Pluto Press.
Esposito, Roberto. 2020. Immunitas. The Protection and Negation of Life. New York: Polity.
Esvelt, Kevin M. and Wang, Harris H. 2013. "Genome-scale engineering for system and synthetic biology". Molecular System Biology, 9: 641.
Eyal, Nir. 2014. Hooked: How to Build Habit-forming Products. New York: Portfolio Penguin.
Feenberg, Andrew. 1999. Questioning Technology. New York: Routledge.
Feldman, Richard, and Conee, Earl. 1985. "Evidentialism". Philosophical Studies, 48, 1: 15-34.
Fernandez-Aranda, Fernando et al. 2012. "Videogames as a complementary therapy tool in mental disorders: PlayMancer, a European multicentre study". Journal of Mental Health, 21, 4: 365-374.
Ferrando, Francesca. 2020. Philosophical Posthumanism. London: Bloomsbury Publishing.
Ferraris, Maurizio. 2020. "L'esplosione della registrazione". In Filosofia del digitale, edited by Luca Taddio and Gabriele Giacomini. Sesto San Giovanni: Mimesis.
Fischer, Matthias, et. al. 2009. "Drosophila RSK negatively regulates bouton number at the neuromuscular junction". Dev. Neurobiol. 69, 4: 212-220.
Floridi, Luciano. 2014. The Fourth Revolution. How the Infosphere is Reshaping Human Reality. Oxford: Oxford University Press.
Flux, John E. C. 1970. "Colour change of mountain hares (Lepus timidus scoticus) in north-east Scotland". J. Zool., 162: 345-358.
Fogel, Alan, et al. 2007. Current problems of Japanese Youth, in Human Development in the Twenty-First Century: Visionary Ideas from Systems Scientists. Cambridge: Cambridge University Press.
Fogg, Brian J. 2005. Persuasive Technology. Burlington: Morgan Kaufmann.
Fotheringham, Michael J., et al. 2000. "Computer use and physical inactivity in young adults: Public health perils and potentials of new information technologies". Annals of Behavioral Medicine, 22: 269-275.
Fournier, Mat. 2018. Biomimétisme: Quand la nature inspire la science. Toulouse: Editions Plume de Carotte.
Francis, Gavin. 2015. Adventures in Human Being. London: Profile Books.

Frankel, Richard, and Krebs, Victor J. 2021. Human Virtuality and Digital Life. London: Routledge.
Frankl, Victor E. 1984. Man's Search for Meaning: An Introduction to Logotherapy. New York: Simon & Schuster.
Franssen, Maarten. 2009. Philosophy of Technology. Stanford Encyclopedia of Philosophy. Stanford CA: Stanford University.
Fredkin, Edward. 1990. "Digital Mechanism: An Informational Process Based on Reversible Universal Cellular Automata". Physica D: Nonlinear Phenomena. 45: 254-270.
Fuchs, Thomas. 2005. "Overcoming Dualism". Philosophy, Psychiatry, & Psychology, 12, 2: 115-117.
Fukuyama, Francis. 2003. Our Posthuman Future: Consequences of the Biotechnology Revolution. New York: Farrar Straus and Giroux.
Gackenbach, Jayne. 2007. Psychology and the Internet. Cambridge: Academic Press.
Gallagher, Shaun, and Zahavi, Dan. 2008. The phenomenological Mind. New York: Routledge.
Garakani, Amir et al. 2020. "Pharmacotherapy of Anxiety Disorder: Current and Emerging Treatment Options". Front Psychiatry, 11: 595584.
Garcia, Tristan. 2010. Form and Object. Paris: Presses Universitaires De France.
Garin, Eugenio. 1958. L'umanesimo italiano. Bari: Laterza.
Garin, Eugenio. 2008. L'uomo del Rinascimento. Bari: Laterza.
Garland, E. Jane. 1998. "Pharmacotherapy of adolescent attention deficit hyperactivity disorder: challenges, choices and caveats". J Psychopharmacol, 12, 4: 385-395.
Gawdat, Mo. 2021. Scary Smart. London: Bluebird Publishing.
Gehlen, Arnold. 1988. Man: His Nature and Place in the World. New York: Columbia University Press.
Gehlen, Arnold. 1988. Man: His Nature and Place in the World. New York: Columbia University Press.
Gerbner, George. 1998. "Cultivation Analysis: An Overview". Mass Communication and Society, 1: 175-194.
Gibbs, Raymond W. Jr. 2006. "Metaphor Interpretation as Embodied Simulation". Mind & Language, 6: 434-458.
Gibson, James J. 1979. The Ecological Approach to Visual Perception. Boston: Houghton Mifflin.
Gibson, Kathleen R., and Ingold, Tim. 1993. Tools, language and cognition in human evolution. Cambridge: Cambridge University Press.

Gilbert, Patrick, and Teglborg, Ann-Charlotte. 2023. Human-technology symbiosis: digital transformation and employee empowerment renewing each other. March 7th 2023, LSE, https://blogs.lse.ac.uk/businessreview/2023/03/07/human-technology-symbiosis-digital-transformation-and-employee-empowerment-renewing-each-other/

Gilbert, Paul. 2017. Human Nature and Suffering. New York: Routledge.

Gilbert, Scott and Epel, David. 2015. Ecological Developmental Biology: The Environmental Regulation of Development, Health and Evolution. Oxford: Oxford University Press.

Gilbert, Scott. 2001. "Ecological Developmental Biology". Developmental Biology, 233, 1: 1-12.

Glezen, W. Paul, and Alpers, Michael. 1999. "Maternal Immunization". Clin. Infect. Dis., 28, 2: 219-224.

Godfrey-Smith, Peter. 2020. Metazoa. Animal Minds and the Birth of Consciousness. Glasgow: William Collins.

Goertzel, Ben. 2007. "Human-level artificial general intelligence and the possibility of a technological singularity: a reaction to Ray Kurzweil's The Singularity Is Near, and McDermott's critique of Kurzweil". Artificial Intelligence, 171, 18: 1161–1173.

Goffman, Erwin. 2017. Interaction Ritual. London: Routledge.

Goldberg, Sanford C. 2016. The Brain in a Vat. Cambridge: Cambridge University Press.

Goodall, Jane. 1990. Through a Window. Boston: Houghton Mifflin Harcourt.

Goudsblom, Johan. 1992. Fire and Civilization. London: Penguin.

Gould, Stephen Jay, and Richard C., Lewontin. 1979. "The spandrels of San Marco and the Panglossian paradigm: a critique of the adaptationist programme". Proceedings of the Royal Society B: Biological Sciences, 205: 581-598.

Gould, Stephen Jay, and Vrba, Elisabeth. 1982. "Exaptation-A Missing Term in the Science of Form." Paleobiology 8, 1: 4-15.

Gould, Stephen Jay. 1977. Ontogeny and phylogeny. Cambridge: Belknap Press of Harvard University Press.

Gould, Stephen Jay. 1989. Wonderful Life. New York: W.W Norton & Company.

Grasso, Donato. 2018. Il formicaio intelligente. Bologna: Zanichelli.

Gray, Chris H. edited by. 1995. Cyborg. Handbook. London: Routledge.

Greenfield, Susan. 2015. Mind Change. How Digital Technology Is Living Their Mark on Our Brains. New York: Random House.

Greguric, Ivana. 2021. Philosophical Issues of Human Cyborgization and the Necessity of Prolegomena on Cyborg Ethics. Hershey: IGI Global.

Guthrie, William Keith Chambers. 1969. A History of Greek Philosophy. Cambridge: Cambridge University Press.

Guzzetti, Barbara J. 2000. "Learning counter-intuitive science concepts: What have we learned from over a decade of research?" Reading & Writing Quarterly: Overcoming Learning Difficulties, 16, 2: 89-98.

Habermas, Jürgen. 2001. On the pragmatics of social interaction. Cambridge: MIT Press.

Haddon, Leslie. 2000. "Social exclusion and information and communication technologies: lesson from studies of single parents and the young elderly". New Media and Society, 2, 4: 387-406.

Han, D.H., et al. 2009. "The effect of methylphenidate on Internet videogame play in children with attention-deficit/hyperactivity disorder". Comprehensive Psychiatry, 50, 3: 251-256.

Harari, Yuval N. 2011. From Animal into Gods. Israel: Kinneret Zmora-Bitan Dvir.

Haraway, Donna J. 2016. Staying with the Trouble, Making Kin in Chthulucene. Durham: Duke University Press.

Haraway, Donna Jeanne. 2016. A Cyborg Manifesto. Minneapolis: University of Minnesota Press.

Harris, Aisha. 2013. "Catfish meaning and definition: term for online hoaxes has a surprisingly long history". Slate, 18 January 2013.

Hashimoto, Teruo, et al. 2013. "Hand before foot? Cortical somatotopy suggests manual dexterity is primitive and evolved independently of bipedalism". Philosophical Transactions of the Royal Society B, 368.

Heath, Christian, Knoblauch, Hubert and Luff, Paul. 2003. "Technology and social interaction: the emergence of 'workplace studies'". The British Journal of Sociology, 51: 299-320.

Hebb, Donald O. 1949. The organization of behavior. A Neuropsychological Theory. New York: Wiley.

Heesy, Christopher H., and Margaret I., Hall. 2010. "The Nocturnal Bottleneck and the Evolution of Mammalian Vision". Brain, Behavior and Evolution. 75:195-203.

Heidegger, Martin. 1959. An introduction to metaphysics. New Haven: Yale University Press.

Heidegger, Martin. 2001. The Fundamental Concepts of Metaphysics. Bloomington: Indiana University Press.

Heil, John. 2003. From an Ontological Point of View. Oxford: Oxford University Press.

Heinemann, Isaak in Pauly-Wissowa, Realencyclopädie der Classischen Altertumswissenschaft, Suppl. V, Stuttgart, 1931.

Hennige, Anita M., et al. 2003. "Upregulation of insulin receptor substrate 2 in pancreatic ß cells prevents diabetes". The Journal of Clinical Investigation, 112, 10: 1521-1532.

Henry, Michel. 1963. L'Essence de la manifestation. Paris: PUF.

Henry, Michel. 2008. Material phenomenology. New York: Fordham University Press.

Herder, Johann Gottfried. 1869. Ideen zur Geschichte der Menschheit. Amsterdam: F.A. Brockhaus.

Hersent, Oliver, et al. 2012. The Internet of Things: Key Applications and Protocols. New York: Wiley.

Higham, Tom. 2021. The World Before Us: How Science Is Revealing a New Story of Our Human Origins. New York: Viking.

Hilton, Julian. 1993. "Theatricality and Technology: Pygmalion and the Myth of the Intelligent Machine". In New Directions in Theatre. New Directions in Theatre, edited by Julian Hilton, 156-175. London: Palgrave.

Hochberg, Leigh R., et al. 2012. Reach and grasp by people with tetraplegia using a neurally controlled robotic arm. "Nature", 485: 372-375.

Hockfield, Susan. 2019. The Age of Living Machines. New York: W.W. Norton & Company.

Horton, James and Priest, Nicholas. 2018. "Silicon Valley's for immortality-and its worrying sacrifices". The Conversation, 101405.

Howard-Jones, Paul. 2011. "The impact of digital technologies on human wellbeing: Evidence from the sciences of mind and brain". Nominet Trust, https://www.thechildrensmediafoundation.org/wp-content/uploads/2014/02/Howard-Jones-2011-impact-digital-technologies-on-wellbeing-copy1.pdf

Hoy, Matthew B. 2018. "Alexa, Siri, Cortana, and More: An Introduction to Voice Assistants". Medical Reference Services Quarterly, 37, 1: 81–88.

Hughes, Howard C. 2001. Sensory Exotica. Bradford: Bradford Book.

Hughes, James. 2004. Citizen Cyborg. Cambridge: Westview Press.

Illich, Ivan. 1973. Tools for conviviality. New York: Harper & Row.

Illmensee, Karl and Levanduski, Mike. 2010. "Embryo splitting". Middle East Fertility Society Journal, 15, 2: 57-63.

Innis, Harold A. 2004. Changing Concepts of Time. Lanham: Rowman & Littlefield Publisher.

Irrgang, Bernhard. 2021. Introduzione alla filosofia della tecnica. Roma. Inschibboleth.
Johanson, Donald, and Edey, Maitland A. 1981. Lucy the Beginnings of Humankind. London: Granada.
Johnson-Laird, Philip N. 1988. The Computer and the Mind. London: William Collins & Co.
Johnson, Steven. 2001. Emergence. New York: Penguin.
Jonas, Hans. 1984. The Imperative of Responsibility: In Search of an Ethics for the Technological Age. Chicago: University of Chicago Press.
Jünger, Ernst. 1932. Der Arbeiter. Hamburg: Hanseatische.
Kamakura, Masaki. 2011. "Royalactin induces queen differentiation in honeybees". Nature, 473: 478-483.
Kardaras, Nicholas. 2016. Glow Kids: How Screen Addiction Is Hijacking Our Kids – and How to Break the Trance. New York: St Martin's Press.
Kauffman, Stuart. 2019. A World beyond Physics. Oxford: Oxford University Press.
Keith, Thomas. 1983. Man and the Natural World. London: Penguin Books.
Kelly, Kevin. 1994. Out of Control. Reading: Addison-Wesley Publishing Company.
Kelly, Kevin. 2011. What Technology Wants. New York: Penguin Books.
Kerskens, Christian M., and Pérez, David L. 2022. "Experimental indications of non-classical brain functions". J. Phys. Commun, 6.
Keverne, Barry, et al. 2005. The Science of Well-Being. Oxford: Oxford University Press.
Kim, Chan Min, et al. 2012. "Teacher Beliefs and technology integration". Teaching and Teacher Education, 29, 1: 76-85.
Kim, Eun-A, and Kang, Seong-Kyu. 2013. "Historical review of the List of Occupational Diseases recommended by the International Labour organization (ILO)". Ann Occup Environ Med., 25: 14.
Kirsh, David. 1995. "The intelligent use of space". Artificial Intelligence, 73, 1-2: 31-68.
Klüwer, Tina. 2011. "From Chatbots to Dialog Systems". In Conversational Agents and Natural Language Interaction: Techniques and Effective Practices, edited by Diana Perez-Marin and Ismael Pascual-Nieto, 1-22. Hershey: Information Science Reference.
Koepp, Matthias J., et al. 1998. "Evidence for striatal dopamine release during a videogame". Nature, 393, 6682: 266-267.
Koffka, Kurt. 1935. Principles of Gestalt Psychology. San Diego: Harcourt.
Konrath, Sara H., et al. 2011. "Changes in dispositional empathy in American college students over time: a meta-analysis". Personality and Social Psychology Review, 15, 2: 180-198.

Koob, George F. and Le Moal, Michel. 2006. Neurobiology of Addiction. Cambridge: Academic Press.
Kortland, Adriaan, and Kooij, M. 1963. "Protohominid Behaviour in Primates". Symposium of the Zoological Society of London, 10: 61-88.
Kringelbach, Morten L., and Berridge, Kent C. 2009. "Toward a Functional Neuroanatomy of Pleasure and Happiness". Trends Cognitive Sciences, 13: 479-487.
Kristeller, Paul O. 1943. Il pensiero filosofico di Marsilio Ficino. Firenze: Le Lettere.
Kristinsdottir, Kolbrun H., et al. 2021. Narcissism and Social Media: The Role of Communal Narcissism. Int. J. Environ. Res. Public Health, 18, 19: 10106.
Kuhn, Thomas. 1957. The Copernican revolution. Cambridge, MA: Harvard University Press.
Kunstler, Barton. 1998. Per una futurologia del sé: come si ridefinisce il concetto di 'identità' in un mondo ipertecnologico. Chicago, 20 July. https://docuver.se/mirrors/www.mediamente.rai.it/mmold/home/bibliote/intervis/k/kunstler.htm.html
Kurzweil, Ray. 1999. The Age of Spiritual Machines: When Computers Exceed Human Intelligence, New York: Penguin Books
Kurzweil, Ray. 2012. How to create a mind: The secret of human thought revealed. New York: Penguin.
Kuss, Daria J., and Griffiths, Mark D. 2012. "Internet gaming addiction: A systematic review of empirical research". International Journal of Mental Health and Addiction, 10, 2: 278-296.
Lacour, Bernard, and Belon, Jean-Paul. 2016. Physiologie humaine. Issy-les-Moulineaux Cedex: Elsevier Masson.
Laland, Kevin N., et al. 1999. "Evolutionary consequences of niche construction and their implications for ecology". Proceedings of the National Academy of Sciences of the Usa, 96: 10242-10247.
Langton, Chris G. 1990. "Computation at the edge of chaos: Phase transitions and emergent computation". Physica D: Nonlinear Phenomena, 42, 1-3: 12-37.
Lanier, Jaron. 2011. You Are Not a Gadget. New York: Penguin.
Lanier, Jaron. 2013. Who Owns the Future?. New York: Simon & Schuster.
Lapidot-Lefler, Noam, and Barak, Azy. 2012. "Effects of anonymity, invisibility, and lack of eye-contact on toxic online disinhibition". Computers in Human Behavior, 28, 2: 434–443.
Larsson, Matz. 2013. "The optic chiasm: a turning point in the evolution of eye/hand coordination". Frontiers in Zoology, 10, 41.

Lasch, Christopher. 1979. The Culture of Narcissism. New York: W.W. Norton & Company.
Lawson, Clive. 2018. Technology and Isolation. Cambridge: Cambridge University Press.
Leakey, Richard. 1994. The Origin of Humankind. New York: BasicBooks.
Leary, Timothy et al. 1964. The Psychedelic Experience: A Manual Based on the Tibetan Book of the Dead. New York: University Press.
LeCun, Yann, et al. 2015. "Deep learning". Nature, 521: 436-444.
LeDoux, Joseph. 1996. The Emotional Brain: The Mysterious Underpinnings of Emotional Life. New York: Simon and Schuster.
Lee, Sojeong, et al. 2015. "Head flexion angle while using a smartphone". Ergonomics, 58:2, 220-226.
LeMoyne, Robert. 2016. Advances for Prosthetic Technology. Japan: Springer.
Leonhard, Gerd. 2016. Technology vs Humanity. London: Fast Future Publishing.
Lerner, Richard M. 1984. On the Nature of Human Plasticity. Cambridge: Cambridge University Press.
Leroi-Gourhan, André. 1945. Milieu et techniques. Évolution et techniques. II. Paris: Albin Michel.
Lévi-Strauss, Claude. 1969. The raw and the cooked. New York: Harper & Row.
Levinas, Emmanuel. 2000. Entre Nous: On Thinking-Of-The-Other. New York: Columbia University Press.
Levins, Richard. 1963. "Theory of fitness in a heterogeneous environment. Developmental flexibility and niche selection". American Naturalist, 47: 75-90.
Lévy, Pierre. 1990. Les Technologies de l'intelligence. L'avenir de la pensée à l'ère informatique. Paris: La Découverte.
Lewis, David. 1954. "Gene-environment interaction: a relationship between dominance, heterosis, phenotypic stability and variability". Heredity, 8: 333-356.
Lewontin, Richard. 1983. "The organism as the subject and object of evolution". Scientia, 118: 63-82.
Libes, Susan M. 2015. "Blue planet: The role of the oceans in nutrient cycling, maintain the atmosphere system, and modulating climate change". In Routledge Handbook of Ocean Resources and Management, edited by Hance D. Smith, Juan Luis Suárez de Vivero and Tundi S. Agardy, 89-108. New York: Routledge.
Lichtenberg, Joseph D. 1989. Psychoanalysis and Motivation. Hillsdale: The Analitytic Press.

Lieberman, Daniel, et al. 2004. "Effects of Food Processing on Masticatory Strain and Craniofacial Growth in a Retrognathic Face". Journal of Human Evolution, 46, 6: 655-677.

Linn, Susan. 2004. Consuming kids: the hostile takeover of childhood. New York: New Press.

Linnet, Jakob, et al. 2010. "Dopamine release in ventral striatum of pathological gamblers losing money". Acta Psychiatrica Scandinavica, 112, 4: 326-333.

Loader, Brian D. 2007. Young Citizens in the Digital Age: Political Engagement, Young People and New Media. New York: Routledge.

Loh, Kep Kee, and, Kanai, Ryota. 2014. "Higher Media Multi-Tasking Activity Is Associated with Smaller Gray-Matter Density in the Anterior Cingulate Cortex". Plose One, 9, 9. https://journals.plos.org/plosone/article/file?id=10.1371/journal.pone.0106698&type=printable

Lombo, José A., and Giménez-Amava, José M. 2014. "The unity and the stability of human behavior. An interdisciplinary approach to habits between philosophy and neuroscience". Front. Hum. Neurosci., 8: 607.

Longo, Giuseppe O. 2003. Il Simbionte. Prove di umanità futura. Milano: Booklet.

Longo, Giuseppe O., and Vaccaro, Andrea. 2013. Bit Bang. La nascita della filosofia digitale. Milano: Apogeo.

Lorenz, Edward N. 1963. "Deterministic non-periodic flow". Journal of Atmospheric Science. 20: 130-141.

Lorenz, Konrad. 1943. "Die angeborenen Formen möglicher Erfahrung". Ethology, 5, 2: 235-409.

Lorenz, Konrad. 1973. Die Rückseite des Spiegels. München: Piper.

Lotto, Beau. 2017. Deviate: The Science of Seeing Differently. London: Hachette Books.

MacNamara, Jim. 2017. Evaluating Public Communication. London: Routledge.

Macrì, Teresa. 1996. Il corpo postorganico. Genova: Costa & Nolan.

Maguire, Eleanor A. et al. 2000. "Navigation-related structural change in the hippocampi of taxi drivers". Proceedings of the National Academy of Science, 97, 8: 4398-4403.

Manetti, Giannozzo. 1975. De dignitate et excellentia hominis. Padova: Editrice Antenore.

Marchesini, Roberto, and Celentano, Marco. 2021. Critical Ethology and Post-Anthropocentric Ethics. Cham: Springer International Publishing.

Marchesini, Roberto. 1996. Il concetto di soglia. Roma: Theoria.

Marchesini, Roberto. 2002. Post-human. Torino: Bollati Boringhieri.

Marchesini, Roberto. 2009. Il tramonto dell'uomo. Bari: Dedalo.
Marchesini, Roberto. 2017. Over the Human. Cham: Springer.
Marchesini, Roberto. 2017. Tecnosfera. Proiezioni per un futuro postumano. Roma: Castelvecchi.
Marchesini, Roberto. 2018. "Animal Being Means Desiring: Subjectivity, Singularity, Diversity in Post-Human Life". In Eco-Phenomenology: Life, Human Life, Post-Human Life in the Harmony of the Cosmos, Analecta Husserliana (The Yearbook of Phenomenological Research), edited by William S. Smith, Jadwiga S. Smith, and Daniela Verducci, 375-385. Heidelberg: Springer.
Marchesini, Roberto. 2021. Le radici del desiderio. Bologna: Apeiron Editoria e Comunicazione.
Marchesini, Roberto. 2021. The Virus Paradigm. Cambridge: Cambridge University Press.
Marchesini, Roberto. 2022. The Creative animal. New York: Palgrave Macmillan.
Marfia, Gustavo, and Matteucci, Giovanni. 2018. Studi di Estetica. Sesto San Giovanni: Mimesis.
Margulis, Lynn. 1970. Origin of Eukaryotic Cells. New Haven: Yale University Press.
Margulis, Lynn. 1991. Symbiosis as a Source of Evolutionary Innovation: Speciation and Morphogenesis. Cambridge: MIT Press.
Margulis, Lynn. 1998. Symbiotic planet: a new look at evolution. New York: Basic Books.
Marion, Jean-Luc and Tardivel, Emilie. 2018. Fenomenologia del dono. Brescia: Morcelliana.
Maslow, Abraham H. 1943. "A theory of human motivation". Psychological Review. 50: 370–396.
Mather, Kenneth. 1953. "Genetical control of stability in development". Heredity, 7: 297-336.
Mayr, Ernst. 1963. Animal species and evolution. Cambridge, MA: Belknap Press.
McCarthy, John, et. al. 2006. "A Proposal for the Dartmouth Summer Research Project on Artificial Intelligence", August 31, 1955. "AI Magazine", 27, 4: 12.
McClelland, James, Rumelhart. David E., and Hinton, Geoffrey E. 1986. "The appeal of PDP". In Parallel distributed processing, edited by James McClelland and David Rumelhart. Cambridge: MIT Press.
McCulloch, Warren, and Pitts, Walter. 1943. "A logical calculus of the ideas immanent in nervous activity". The bulletin of mathematical biophysics, 5: 115-133.

McCutcheon John P. and Nancy A. Moran. 2011. "Extreme genome reduction in symbiotic bacteria". Nature Rev. Microbiol., 10: 13-26.
McFall-Ngai, Margaret, et al. 2013. "Animals in a bacterial world, a new imperative for the life sciences". Proc Natl Acad Sci USA, 110, 9: 3229-3236.
McKay, Jill A., and Mathers, John C. 2011. "Diet induced epigenetic changes and their implications for healt". Acta Physiologica, 202, 2: 103-118.
McLuhan, Marshall. 1964. Understanding Media: The Extension of Man. New York: McGraw Hill.
McMullan, Thomas. 2015. "What does the panopticon mean in the age of digital surveillance?". The Guardian. July 23 2015.
McNamee, P. et al. 2021. "Social media use and emotional and behavioural outcomes in adolescence: Evidence from British longitudinal data". Economics & Human Biology, 41: 100992.
McQuail, Denis, and Sven, Windahl. 1993. Communication Models. London: Routledge.
Merleau-Ponty, Maurice. 1964. L'œil et l'esprit. Paris: Editions Gallimard.
Merleau-Ponty, Maurice. 2003. Nature. Evanston: Northwestern University Press.
Metcalfe, Janet, and Shimamura, Arthur P. 1994. Metacognition: knowing about knowing. Cambridge: MIT Press.
Miller, George, et al. 2013. Plans and the Structure of Behavior. Eastford: Martino Fine Books.
Minelli, Alessandro, and Fusco, Giuseppe. 2004. "Evo-devo perspectives on segmentation: model organisms, and beyond". Trends Ecol. Evol., 19, 8: 423-429.
Minsky, Marvin, and Seymour, Papert. 1972. Artificial Intelligence. Eugene OR: University of Oregon Press.
Mitchell, Barbara A. 2006. The Boomerang Age. Transaction to Adulthood in Families. New York: Routledge
Mitchell, Melania. 2019. Artificial Intelligence. New Orleans: Pelican.
Mitchell, Melanie. 2019. Artificial intelligence: A guide for thinking humans. New York: Ferrar, Straus and Giroux.
Mitchell, Melanie. 2019. Artificial Intelligence. New York: Ferrar Straus and Giroux.
Moczek, Armin P. 2015. "Re-evaluating the environment in developmental evolution". Front. Ecol. Evol., 3: 1.
Moffett, Mark W. 2018. The Human Swarm: How Our Societies Arise, Thrive, and Fall. Bisignano: Apollo.
Molina, Patricia E. 2018. Endocrine Physiology. New York: McGraw Hill.

Montagu, Ashley. 1956. "Neoteny and the evolution of human mind". Explorations, 5.

Monteanu, Teona, et al. 2018. "Light-dependent pathways for dopaminergic amacrine cell development and function". Elife, 7: e39866.

Moravec, Hans. 1998. Robot. Mere Machine to Transcendent Mind. Oxford: Oxford University Press.

More, Max. 1995. The Diachronic Self: Identity, Continuity, Transformation. Usa: A. Bell & Howell.

Morin, Edgar. 1974. L'Unité de l'homme: essais de discussion. Paris: Seuil.

Morozov, Evgeny. 2011. The Net Delusion. New York: Public Affairs.

Morris, Desmond. 1967. The Naked Ape. New York: Dell Publishing.

Nagy, Peter and Koles, Bernadett. 2014. "The digital transformation of human identity". Convergence, 20, 3: 276-292.

Narayan, Jyoti, et al. 2023. "Elon Musk and others urge Al pause, citing "risks to society". Reuters, https://www.reuters.com/technology/musk-experts-urge-pause-training-ai-systems-that-can-outperform-gpt-4-2023-03-29/

Needleman, Herbert L. 1999. "History of lead poisoning in the world". International Conference on lead Poisoning Prevention and Treatment. Bangalore, February 8-10, 1999.

Nesi, Jacqueline. 2020. "The impact of social media on youth mental health: challenges and opportunities". North Carolina Medical Journal, 81, 2: 116-121.

Nestler, Eric J. 2014. "Epigenetic mechanisms of drug addiction". Neuropharmacology. 76: 259-268

Neupane, Sunil, et al. 2017. "Text Neck Syndrome-Systematic Review". Imperial Journal of Interdisciplinary Research, 3, 7:141-148.

Nida-Rümelin, Julian and Weidenfeld, Nathalie. 2018. Digitaler Humanismus. Heine Ethik für das Zeitalter der Künstlichen Intelligence. München/Berlin: Piper.

Nietzsche, Friedrich. 1917. Thus Spoke Zarathustra. New York, Modern Library.

Nijhout, Frederik. 1999. "Control Mechanisms of Polyphenic Development in Insects: In polyphenic development, environmental factors alter some aspects of development in an orderly and predictable way". BioScience, 49, 3: 181-192.

Nippert-Eng, Christena. 1996. "Calendars and Keys: The classification of "home" and "work". Sociological Forum, 11: 563-582.

Noble, David F. 1997. The Religion of Technology. The Divinity of Man and the Spirit of Invention. New York: Knopf.

Noë, Alva. 2009. Out of the Head. Why You Are Not Your Brain. Cambridge: The MIT Press.
Norman, David. 1994. Prehistoric Life. Hoboken: John Wiley & Sons.
Norman, Donald. 1994. Things That Make Us Smart: Defending Human Attributes in the Age of the Machine. Cambridge: The MIT Press.
Nuismer, Scott L. 2017. Introduction to coevolutionary theory. New York: Macmillan Higher Education.
Nussey, Daniel H., et al. 2005. "Selection on hereditable phenotypic plasticity in a wild bird population". Science, 310: 304-306.
Nutt, John. 1915. Disease and Deformities of the Foot. New York: E.B. Treat.
O'Connel, Mark. 2018. To be a Machine. London: Granta Books.
O'Connor, Timothy, and Wong, Hong Yu. 2012. Emergent Properties. In Stanford Encyclopedia of Philosophy, edited by Edward N. Zalta. Summer 2015 edition.
O'Doherty, Joseph E., et al. 2011. "Active tactile exploration using a brain-machine-brain interface". Nature, 479: 228-231.
Odling-Smee, John F. 1988. "Niche Constructing Phenotypes". In the Role of Behaviour in Evolution, edited by Henry C. Plotkin. Cambridge: MIT Press.
Odling-Smee, John F. et al. 2017. Niche construction, "Am. Nat.", 147: 641-648.
Odum, Eugen P., and Barrett, Gary W. 2005. Fundamentals of Ecology. Belmont, CA: Thomson Brooks-Cole.
Olsen, Jan K. B., and Christiansen, Frederik V. 2009. "Technology and Science epistemology, rationality and the empirical turn". Synthese, 168: 313-318.
Ophir, Nass, et al. 2009. Cognitive control in media multitaskers. "Proceedings of the National Academy of Sciences of the United States of America". 106, 37: 15583-15587.
Owen, John B., et. al. 2001. Animal Models- Disorders of Eating Behaviour and Body Composition. Norwell-Mass: Kluwer Academic Publishers.
Özdemir, Vural et al. 2021. "Thanatechnology and the Living Dead: New Concepts in Digital Transformation and Human-Computer Interaction". OMICS A Journal of Integrative Biology, 25, 7: 401-407.
Paaby, Annalise B., and Rockman, Matthew V. 2016. "The many faces of pleiotropy". Trends in Genetics. 29, 2: 66–73.
Palmer, Sue. 2007. Toxic childhood: How the modern world is damaging our children and what we can do about it. London: Orion.

Panksepp, Jaak, and Biven, Lucy. 2012. The Archaeology of Mind: Neuroevolutionary Origins of Human Emotions. New York: W.W. Norton & Company.

Pariser, Eli. 2011. The Filter Bubble. New York: Penguin.

Patocka, Jan. 2016. Le monde naturel comme problème philosophique. Paris: Vrin.

Pechenik, Jan A., et al. 1998. "Metamorphosis is not a new beginning". BioScience, 48: 901-910.

Pellegrino, Giuseppina. 2006. "Ubiquity and Pervasivity: On the Technological Mediation of (Mobile) Everyday Life. In Social Informatics: An Information Society for all?". In Remembrance of Rob Kling. Edited by Berleur J., 133-144. IFIP International Federation for Information Processing, 223. Boston M: Springer.

Perlis, Roy H., et al. 2021. "Association between social media use and self-reported symptoms of depression in US adults". JAMA Network Open, 4, 11.

Perren, Sonja, and Gutzwiller-Helfenfinger, Evelin. 2012. "Cyberbullying and Traditional bullying in adolescence: Differential roles of moral disengagement, moral emotions, and moral values". European Journal of Developmental Psychology, 9, 2: 195-209.

Petrosyan, Maria I. 1972. Humanism: Its Philosophical, Ethical and Sociological Aspect. Moscow: Progress.

Pettitt, Paul. 2022. Homo Sapiens Rediscoveder. London: Thames & Hudson.

Pfister, Cuno. 2011. Getting Started with the Internet of Things. Sebastopol: O'Reilly Media.

Piaget, Jean. 1970. Psychologie et épistémologie. Paris: Gonthier.

Piazza, Pier Vincenzo. 2019. Homo Biologicus. Milano: Rizzoli.

Pico della Mirandola, Giovanni. 1942. Oratio de homini dignitate. Firenze: Vallecchi.

Pierson, John. 2002. Tackling Social Exclusion. New York: Routledge.

Pigliucci, Massimo, and Müller, Gerd B. 2010. Evolution. The Extended Synthesis. Cambridge: The MIT Press.

Pigliucci, Massimo. 2001. Phenotypic Plasticity: Beyond Nature and Nurture. Baltimore: Johns Hopkins University Press.

Plessner, Helmuth. 1928. Die Stufen des Organischen und der Mensch. Einleitung in die philosophische Anthropologie. Berlin: The Gruyter.

Plessner, Helmuth. 2019. Levels of Organic Life and the Human. New York: Fordham University Press.

Ponti, Crystal. 2016. Can the Future of Brain Preservation Help Us Live Forever? https://medium.com/the-establishment/can-the-future-of-brain-preservation-help-us-live-forever-def5f2d0ed52

Porges, Stephen W. 2011. The Polyvagal Theory. New York: W.W. Norton & Company.

Portelli, Andrew, and Reid, Susan A. 2018. "Cervical Proprioception in a Young Population Who Spend Long Periods on Mobile Devices: A 2-Group Comparative Observational Study". Journal of Manipulative and Physiological Therapeutics, 41, 2: 123-128.

Portmann, Adolf. 1945. A Thinker of Self-Expressive Life. Cham: Springer.

Postman, Neil. 1994. Das Technopol: die Macht der Technologien und die Entmündigung der Gesellschaft. Frankfurt: S. Fisher.

Prakash, Pal C., et al. 2016. "Epidemiology of Knee Osteoarthritis in India and Related Factors". Indian Journal of Orthopaedics, 50, 5: 518-522.

Prensky, Marc. 2001. "Digital natives, Digital immigrants: Part 1". On the Horizon, 9: 1-6.

Prigogine, Ilya, and Stengers, Isabelle. 1979. La Nouvelle Alliance, Métamorphose de la science. Paris: Édition Gallimard.

Prigogine, Ilya. 1967. "Introduction to thermodynamics of irreversible processes". Interscience New York, 3: 147.

Prutner, Wiebke. 2013. "Hazard and risk assessment of teratogenic chemicals under REACH". Methods Mol. Biol., 947: 517-543.

Przybylski, Andrew K., et al. 2013. Motivational, emotional, and behavioral correlates of fear of missing out. 29, 4: 1841-1848.

Pujadas, Elisabet, and Feinberg, Andrew P. 2012. "Regulated noise in the epigenetic landscape of development and disease". Cell., 148, 6: 1123-31.

Pulcini, Elena. 2012. Care of The World. New York: Springer.

Purcell, Kristen, et al. 2012. How teens do research in the digital world. https://www.pewresearch.org/internet/2012/11/01/how-teens-do-research-in-the-digital-world-2/

Putnam, Hilary. 1975. "Minds and Machines". In Dimension if Mind, edited by Sidney Hook, 392-415. New York: New York University Press.

Raff, Rudy A., and Kaufman, Thomas C. 1983. Embryos, Genes and Evolution. New York: Macmillan Publishing Co.

Ragnedda, Massimo and Muschert, Glenn W. 2013. The Digital Divide. London: Routledge.

Raskin, Robert, and Terry, Howard. 1988. "A principal-components analysis of the Narcissistic Personality Inventory and further evidence

of its construct validity". Journal of Personality and Social Psychology, 54, 5: 890-902.
Reader, Simon M., and Laland, Kevin N. 2003. Animal Innovation. New York: Oxford University Press.
Remotti, Francesco. 1996. Contro l'identità. Bari: Laterza.
Renner, Michael, edited by. 2013. Vital Sign, The Trends that are Shaping Our Future. Washington: World Watch Institute.
Reynolds, Robert G. 2009. "A Genealogy of Computational Intelligence Books". Idee Computational Intelligence Magazine, 2: 56-61.
Reza Jorjani, Jason. 2020. Prometheism. Budapest: Arktos Media.
Rheingold, Howard. 1992. Virtual Reality. Touchstone: Simon & Schuster.
Ridker, Nader, et al. 2002. "Comparison of C-reactive protein and low-density lipoprotein cholesterol levels in the prediction of first cardiovascular events". N. Engl J. Med. 347: 1557-1565.
Rifkin, Jeremy. 1983. Algeny: A New Word. A New World. New York: Viking.
Rivoltella, Pier Cesare, and Rossi, Pier Giuseppe. 2019. Il corpo e la macchina. Brescia: Morcelliana.
Robert J. Lifton. 1994. The Protean Self: Human Resilience in an Age of Fragmentation. New York: Basic Books, 1993.
Romo, Ranulfo, et al. 2000. "Sensing without Touching: Psychophysical Performance Based on Cortical Microstimulation". Neuron, 26, 1: 273-278.
Roosendaal, Arnold. 2013. Digital Personae and Profiles in Law. The Netherlands: Wolf Legal Publishers.
Rosa, Hartmut. 2010. Alienation and Acceleration. Aarhus: Aarhus University Press.
Rosen, Larry D. 2012. iDisorder: Understanding our obsession with technology and overcoming its hold on us. New York: Macmillan.
Rosenberg, Eugene, et al. 2007. "The Role of Microorganisms in coral health, disease and evolution". Nature Rev. Microbiol., 5: 355-362.
Rosmini, Antonio. 2009. Scritti pedagogici. Stresa: Edizioni Rosminiane Sodalitas.
Rubin, James G., et al. 2011. "Do people with idiopathic environmental intolerance attributed to electromagnetic fields display physiological effects when exposed to electromagnetic fields? A systematic review of provocation studies". Bioelectromagnetics, 32, 8: 593-609.
Rupert, Robert D. 2009. Cognitive System and the Extended Mind. Oxford: Oxford University Press.
Rushkoff, Douglas. 2013. Present Shock. When Everything Happens Now. London: Current.

Rutledge, Thomas, et al. 2008. "Social networks and incident stroke among women whit suspected myocardial ischemia". Psychosomatic Medicine, 70, 3: 282-287.
Sadin, Eric. 2015. La Vie algorithmique. Paris: L'Échappée.
Sahlins, Marshall D. 1972. Stone Age Economics. Chicago: Aldine.
Sakamoto, Kei. et al. 2003. "Akt signaling in skeletal muscle: Regulation by exercise and passive stretch". Am. J. Endocrinol. Metab. 285: E1081-E1088.
Sapolsky, Robert M. 2004. Why Zebras don't Get Ulcers. New York: Henry Holt and Company.
Sax, Boria. 2021. Avian Illuminations. London: Reaktion Books.
Scarmeas, Nikolaos and Stern, Yaakov. 2003. "Cognitive reserve and lifestyle". Journal of Clinical and Experimental Neuropsychology, 25, 5: 625-633.
Schaab, Gloria. 2013. "Incarnation as Emergence: A Transformative Vision of God and the Cosmos". The Heythrop Journal LIV, 54: 631-644.
Scheler, Max. 1955. Vom Umsturz der Werte: Abhandlungen und Aufsätze, edited by Maria Scheler. Bern: Francke.
Scheler, Max. 1970. The nature of sympathy. Oxfordshire: Routledge & K. Paul.
Scheler, Max. 2009. The Human Place in the Cosmos. Evanston: Northwestern University Press.
Schrödinger, Erwin. 1940. What is Life? Cambridge: Cambridge University Press.
Schulkin, Jay. 2004. Allostasis, Homeostasis, and the Costs of Physiological Adaptation. Cambridge: Harvard University Press.
Searle, John. 1995. The Construction of Social Reality. New York: Free Press.
Sha, Peng, and Dong, Xiaoyu. 2021. "Research on adolescents regarding the indirect effect of depression, anxiety, and stress between TikTok use disorder and memory loss". International Journal of Environmental Research and Public Healt, 18, 16: 8820.
Shapiro, Lawrence. 2011. Embodied cognition. New York: Routledge.
Shapiro, Michael D., et al. 2003. "Developmental basis of evolutionary digit loss in the Australian lizard". Hemeergis J. Exp. Zool. 279B: 48-56.
Sherwood, Lauralee, et al. 2012. Animal Physiology: From Genes to Organisms. Boston: Cengage Learning.
Shilling, Chris. 2016. "The Rise of Body Studies and the Embodiment of Society: A Review of the Field". Horizons in Humanities and Social Sciences: An International Refereed Journal, 2, 1: 1-14.

Shirow, Masamune. 1989. The Gost in the Shell. Tokyo: Kodansha Young Magazine Zōkan Kaizokuban.

Siegel, Allan, and Sapru, Hreday N. 2015. Essential Neuroscience. Philadelphia: Lippincott Williams & Wilkins.

Siegel, Daniel J. 1999. The developing mind. Toward a neurobiology of interpersonal experience. New York: Guilford Press.

Sigman, Aric. 2009. "Well connected? The biological implications of "social networking", Biologist, 56, 1: 14-20.

Simondon, Gilbert. 2020. Individuation in Light of Notions of From and Information. Minneapolis: University of Minnesota Press.

Simpson, Stephen J., et al. 2011. "Polyphenism in insects". Corr. Biol. 21, 18: 738-749.

Sisto, Davide. 2018. La morte si fa social. Immortalità, memoria e lutto nell'epoca della cultura digitale. Torino: Bollati Boringhieri.

Sisto, Davide. 2022. Porcospini digitali. Vivere e mai morire online. Torino: Bollati Boringhieri.

Skinner, Michael K., et al., 2010. "Epigenetic transgenerational actions of endocrine disruptors". Reprod. Toxicol., 31: 337-343.

Skues Jason L., et al. 2012. "The effects of personality traits, self-esteem, loneliness, and narcissism on Facebook use amongst university students". Computers in Human Behavior, 28, 6: 2414-2419.

Slettemeås, Dag. 2014. Bruk av smarttelefon og nettbrett i den norske befolkningen. SIFO-survey hurtigstatistikk 2013. Prosjektnotat 2. Oslo: SIFO.

Sloterdijk, Peter. 2011. Bubbles. Spheres Vol I: Microsphereology. Los Angeles: Semiotext(e).

Small, Gary, and Vorgan, Gigi. 2009. El cerebro digital. Cómo las nuevas tecnologías están cambiando nuestra mente. Barcelona: Ediciones Urano.

Smith, Brian C. 1999. "Situatedness/embeddedness". In The MIT encyclopedia of the cognitive sciences, edited by Robert A. Wilson and Frank C. Keil, 769–770. Cambridge: MIT Press.

Smith, Marquard. 2005. The Monograph. London-Cambridge: The MIT Press.

Sofka, Carla, et al. edited by. 2012. Dying, Death, and Grief in an Online Universe. New York: Springer Publishing Company.

Solms, Mark. 2021. The Hidden Spring. A Journey to the Source. London: Profile Books.

Sorgner, Stephan L. 2022. We have always been cyborgs. Digital data, gene technologies, and an ethics of transhumanism. Bristol: Bristol University Press.

Soto, Ana M., et al. 1991. "*p*-Nonylphenol: An Estrogenic Xenobiotic Released from "Modified" Polystyrene". Environ. Health Persp. 92: 167-173.

Sparrow, Bethsy, et al. 2011. "Google Effects on Memory: Cognitive Consequences of Having Information at Our Fingertips". Science, 333, 6043: 776-778.

Sperber, Dan, and Wilson, Deirdre. 1986. Relevance: Communication and Cognition. Oxford: Blackwell's.

Spitzer, Manfred. 2012. Digital Demenz. München: Droemersche.

Sterelny, Kim. 2003. Thought in a hostile world. Oxford: Blackwell Publishing.

Steup, Matthias and Ichikawa, Jonathan Jenkins. 2018. The Analysis of Knowledge. Stanford: Stanford University.

Steven, Jones G. 1997. Virtual Culture. London: Sage Publications.

Stevens, John, et al. 2017. "A Review of Small Screen and Internet Technology-Induced Pathology as a Lifestyle Determinant of Health and Illness". Am J Lifestyle Med., 14, 2: 114-117.

Stiegler, Bernard. 2016. Automatic Society. New York: Wiley.

Stout, Dietrich, et al. 2008. "Neural correlates of Early Stone Age toolmaking: technology, language and cognition in human evolution". Philos. Trans. R. Soc. Lond. B Biol. Sci., 363, 1499: 1939-1949.

Suler, John R. 2015. Psychology of the Digital Age. Cambridge: Cambridge University Press.

Suler, John. 2004. "The online disinhibition effect". Cyberpsychology & Behavior, 7, 3: 321-326.

Swan, Melanie. 2013. "The Quantified Self". Big Data, 1, 2: 85-99.

Swan, Shanna H. 2021. Countdown. New York: Scribner.

Sweetser, Kaye D. 2010. "A Losing Strategy: The Impact of Nondisclosure in Social Media on Relationships". Journal of Public Relations Research, 22: 288-312.

Talwar, Sanjiv K., et al. 2002. "Behavioural neuroscience: Rat navigation guided by remote control". Nature, 417: 37-38.

Tamir, Diana I., and Mitchell, Jason P. 2012. "Disclosing information about the self is intrinsically rewarding". PNAS, 109, 21: 8038-8043.

Tapscott, Don. 1998. Growing up digital: the rise of the net generation. New York: McGraw Hill.

Tapscott, Don. 2009. Grown up Digital: How the Net Generation Is Changing the World. New York: McGraw Hill.

Teperino, Raffaele. 2020. Beyond Our Genes. Cham: Springer.

Terrosi, Roberto. 1997. La filosofia del postumano. Genova: Costa & Nolan.

Textor, Mark. 2017. Brentano's Mind. Oxford: Oxford University Press.
Thagard, Paul. 1996. Mind: introduction to Cognitive Science. Bradford: Bradford Books.
Thelen, Esther, and Smith, Linda B. 1994. A dynamic system approach to the development of cognition and action. Cambridge: The MIT Press.
Thomas, Lewis. 1974. The Lives of a Cell: Notes of a Biology Watcher. New York: Viking.
Thorson, Jennifer L.M., et al. 2021. "Ancestral plastics exposure induces transgenerational disease-specific sperm epigenome-wide association biomarkers". Environmental Epigenetics, 7, 1: 1-13.
Thweatt-Bates, Jeanine. 2016. Cyborg Selves: A theological Anthropology of the Posthuman. London: Routledge.
Tiffon, Céline. 2018. "The Impact of Nutrition and Environmental Epigenetics on Human Healt and Disease". Int. Mol. Sci., 19: 3425.
Tinbergen, Nikolaas. 1965. Animal Behavior. New York: Time Life.
Tinsley, Joyce, et al. 1994. Fluoxetine Abuse. "Rapid Communication", 69, 2: 166-168.
Toffler, Alvin. 1970. Future Shock. New York: Random House.
Tomasello, Michael. 2009. Why We Cooperate. Cambridge: The MIT Press.
Tosun, L.P. and T. Lajunen. 2009. Why do young adults develop a passion for Internet activities? The associations among personality, revealing "true self" on the Internet, and passion for the Internet. "Cyber Psychology & Behavior, 12, 4: 401-406.
Travers Scott, D. 2018. Pathology and Technology. New York: Media and Communication.
Trepte, Sabine, and Reinecke, Leonard. 2013. "The reciprocal effects of social network site use and the disposition for self-disclosure: A longitudinal study". Computers in Human Behavior, 29, 3: 1102-1112.
Turkle, Sherry. 2011. Alone together: Why we expect more from technology and less from each other. New York: Basic Books.
Tversky, Amos, and Kahneman, Daniel. 1974. "Judgment under Uncertainty: Heuristic and Biases". Science, 185, 4157: 1124-1131.
Tymieniecka, Anna Teresa. 2007. "Phenomenology of Life. From the Animal Soul to the Human Mind", Book II, Analecta Husserliana. New York: Springer.
Vaidhyanathan, Siva. 2018. Anti-social Media. Oxford: Oxford University Press.
Varela, Francisco J. et al. 1992. The Embodied Mind. Cambridge: The MIT Press.

Varrela, Juha. 1992. "Dimensional variation of craniofacial structures in relation to changing masticatory-functional demands". Eur. J. Orthod., 14: 31-36

Vatsiou, Alexandra I., et al. 2016. "Detection of selective sweeps in structured population: a comparison of recent methods". Molecular Ecology, 25, 1: 89-103.

Velliste, Meel, et al. 2008. "Cortical control of a prosthetic arm for self-feeding". Nature, 453: 1098-1101.

Veneris, Y. 1990. "Modelling the Transition from the Industrial to the Informational Revolution". Environment and Planning A, 22, 3: 399-416.

Verbeek, Peter Paul. 2015. "Beyond interaction: a short introduction to mediation theory". Interactions, 22, 3: 26-31.

Vinge, Vernor. 1993. "Technological Singularity". Whole Earth Review, 81.

Virilio, Paul. 1995. La vitesse de libération. Paris: Éditions Galilée.

Von Neumann, John. 1958. The Computer and the Brain. New Haven: Yale University Press.

Von Uexküll, Jakob. 1956. Streifzüge durch die Umwelten von Tieren und Menschen: Ein Bilderbuch unsichtbarer Welten. Bedeutungslehre. Hamburg: Rowohlt.

Von Uexküll, Jakob. 1956. Streifzüge durch die Umwelten von Tieren und Menschen: Ein Bilderbuch unsichtbarer Welten. Bedeutungslehre. Hamburg: Rowohlt.

Vygotskij, Lev S. 1962. Thinking and Speaking. Cambridge: The MIT Press.

Wachs, Sebastian, and Write, Michelle F. 2018. "Associations between Bystanders and Perpetrators of Online Hate: The Moderating Role of Toxic Online Disinhibition". International Journal of Environmental Research and Public Health, 15, 2030.

Waddington, Conrad H. 1959. "Evolutionary adaptation". In Evolution after Darwin, edited by Sol Tax, vol. 1, 381-402. Chicago: University Chicago Press.

Wagner, Günther P., and Laubichler, Manfred D. 2004. "Rupert Reiedl and their-synthesis of evolutionary and developmental biology: Body plans and evolvability". J. Exp. Zool. (MDB), 302B: 92-102.

Wallace, Arthur. 2021. Understanding evo-devo. Cambridge: Cambridge University Press.

Wallace, Patricia. 2014. "Internet addiction disorder and youth: There are growing concerns about compulsive online activity and that this could

impede students' performance and social lives". EMBO Rep, 15, 1: 12-16.
Wang, Chuang, et al. 2015. "A theory of social media dependence: Evidence from microblog users". Decision Support System, 69: 40-49.
Warner, Jon, and Sandberg, Anne. 2007. Generational Styles Assessment Facilitator's Guide. Human Resource Development.
Warnke, Ulrich. 2017. Bionische Regeneration: Das Altern aufhalten mit den geheimen Strategien der Natur. München: Arkana.
Warren, Janet M., et al. 2010. "Assessment of physical activity. A review of methodologies with reference to epidemiological research: a report of the exercise physiology section of the European Association of Cardiovascular Prevention and Rehabilitation". European Journal of Preventive Cardiology, 17, 2: 127-139.
Warschauer, Mark. 2004. Technology and Social Inclusion: Rethinking the Digital Divide. Cambridge: MIT Press.
Weaver Ian C. G. et al. 2007. "The Transcription Factor Nerve Growth Factor-Inducible Protein A Mediates Epigenetic Programming: Altering Epigenetic Marks by Immediate-Early Genes". Journal of Neuroscience, 27, 7: 1756-1768.
Weaver, Jeremiah ed. 2016. Animal Welfare. New York: Nova Science Publisher.
Weil, Michelle M. and Rosen, Larry D. 1997. Technostress Coping with Technology @Work @Home @Play. Hoboken: John Wiley & Sons.
West-Eberhard, Mary Jane. 2003. Developmental Plasticity and Evolution. Oxford: Oxford University Press.
Wheeler, Steve. 2012. "Digital literacies for engagement in emerging online cultures". Research Paper Series, 5: 14-25.
Wiener, Norbert. 1948. Cybernetics, or control and communication in the animal and the machine. Cambridge: MIT Press.
Wiley, Keith. 2014. A Taxonomy and Metaphysics of Mind-Uploading. Seattle: Humanity+Press and Alautun Press.
Wilson, Edward O. 1978. On Human Nature. Cambridge: Harvard University Press.
Wilson, Frank R. 1999. The hand: How its use shapes the brain, language, and human culture. New York: Vintage.
Wilson, Robert A., and Clark, Andy. 2009. "How to Situate Cognition: Letting Nature Take its Course". In The Cambridge Handbook of Situated Cognition, edited by Philip Robbins and Murat Aydede, 55-77. New York: Cambridge University Press.
Winnicott, Donald W. 1971. Playing and Reality. London-New York: Tavistock.

Wolff, Julius. 1986. The Law of Bone Remodeling. Heidelberg: Springer-Verlag Berlin.

Wolfram, Stephen. 2002. A New Kind of Science. Champaign Illinois: Wolfram Media.

Wolpert, Lewis. 2000. The Unnatural Nature of Science. Cambridge: Harvard University Press.

Wood Heater C., and Scott, Holly. 2016. "Social media use in adolescence is associated with poor sleep quality, anxiety, depression and low self-esteem". J. Adolesc, 51: 41-49.

World Health Organization. 2018. Global Status Report on Road Safety 2018.

Wrangham, Richard. 2019. The Goodness Paradox: The Strange Relationship Between Virtue and Violence in Human Evolution. London: Profile Books.

Xiao, Jin-Hua. et al. 2013. "Obligate mutualism within a host drives the extreme specialization of a fig wasp genome". Genome Biol., 14, 12: R141.

Xie, Haoran, et al. 2019. "Trends and development in technology-enhanced adaptive/personalized learning: A systematic review of journal publications from 2007 to 2017". Computers & Education, 140.

Yakoubian, Mane K. 2021. "Brains are getting smaller in modern humans". Psypost, November 16.

Yan, Zheng. 2020. "Unprecedented pandemic, unprecedented shift, and unprecedented opportunity". Hum. Behav. & Tech., 2: 110-112.

Yuan, Jie et al. 2020. "Brain signal Changes of sensory cortex according to surface roughness of boneless corsets". Textile Research Journal. 90, 1: 76-90.

Zaliauskiene, Lolita et al. 2016. "Down-Regulation of Cell Surface Receptors Is Modulated by Polar Residues within the Transmembrane Domain". Molecular Biology of the Cell, 11, 8: 2643-2655.

Zeitzoff, Thomas. 2017. "How Social Media is Changing Conflict". Journal of Conflict Resolution, 61, 9: 1970-1991.

Zhao, Shanyang, et al. 2008. "Identity construction on Facebook: Digital empowerment in anchored relationships". Computers in Human Behavior, 24: 1816-1836.

Zhok, Andrea. 2011. Emergentismo. Pisa: Edizioni ETS.